T0141136

TRANSHUMANISM

TRANSHUMANISM

Evolutionary Futurism and the Human Technologies of Utopia

ANDREW PILSCH

University of Minnesota Press
Minneapolis
London

A different version of chapter 2 was published as "Fan Utopias and Self-Help Supermen: Political Utopianism in World War II–Era Science Fiction," *Science Fiction Studies* 41, no. 3 (2014): 524–42.

Published by the University of Minnesota Press
111 Third Avenue South, Suite 290
Minneapolis, MN 55401-2520
http://www.upress.umn.edu

Library of Congress Cataloging-in-Publication Data
Names: Pilsch, Andrew, author.
Title: Transhumanism : evolutionary futurism and the human technologies of utopia / Andrew Pilsch.
Description: Minneapolis : University of Minnesota Press, 2017. | Includes bibliographical references and index.
Identifiers: LCCN 2016039456 (print) | ISBN 978-1-5179-0101-1 (hc) |
ISBN 978-1-5179-0102-8 (pb)
Subjects: LCSH: Humanism. | Philosophical anthropology.
Classification: LCC B821 .P46 2017 (print) | LCC B821 (ebook) | DDC 144—dc23
LC record available at https://lccn.loc.gov/2016039456

UMP LSI

CONTENTS

INTRODUCTION

What Is Transhumanism?

We exist in an age in which human futures have radically changed. As the revolutionary energy flickered out at the end of the psychedelic sixties, an organized movement of technologists, philosophers, and scientists calling itself "transhumanism" began to emerge, especially in Western Europe and the United States. First named in its modern incarnation by FM-2030,[1] transhumanism is an increasingly pervasive movement and an important actant, especially in technology policy and bioethics debates, whose members seek, broadly, to hack the human biocomputer to extend life, increase welfare, and enhance the human condition. As Max More, transhuman philosopher and advocate for radical life extension, defined in 1990, "transhumanism is a class of philosophies of life that seek the continuation and acceleration of the evolution of intelligent life beyond its currently human form and human limitations by means of science and technology, guided by life-promoting principles and values."[2] More's revised 1996 version of this definition also clarifies that the movement seeks to use "a rational philosophy and values system" to recognize and anticipate "the radical alterations in the nature and possibilities of our lives resulting from various sciences and technologies such as neuroscience and neuropharmacology, life extension, nanotechnology, artificial ultraintelligence, and space habitation."[3]

Since the 1970s, transhumanism has been gaining speed and influence, particularly among people who work with computers in places such as California's Silicon Valley or New York's Silicon Alley, as the dizzying pace of information technology appears to give increasing credence to the ideas futurists such as FM-2030 and Max More have been documenting for decades. Inspired by theories of ever-outward technological expansion—such as Moore's Law, which states that computing performance doubles every eighteen months, or George Dyson's observation that "global production of optical fiber reached Mach 20 (15,000 miles per hour) in 2011, barely keeping up with demand"—transhuman philosophy articulates a stance toward communication technology that sees such massive, planetary expansion of humanity's technological reach as suggesting coming mutations in the basic nature of the human condition.[4] Transhumanism, Max More suggests, comes after humanism, which More suggests was "a major step in the right direction" but one that "contains too many outdated values and ideas."[5] Transhumanists, according to More, seek to remix the legacy of humanist thought from a "philosophy of life that rejects deities, faith, and worship, instead basing a view of values and meaningfulness on the nature and potentials of humans within a rational and scientific framework" into one that embraces and anticipates the radical changes brought about by planetary communication technologies and radical technologies of the body.[6]

In More's articulation of the movement's philosophy, which has since become *the* reference point for self-definition within the movement, transhumanism represents a new form of the future that emerges in response to these planetary changes. Most of the futures we are used to in popular culture borrow from classical narratives of the American Golden Age of science fiction (SF) and signal their futurity through some kind of new technology external to our bodies (think of the various technological marvels of the *USS Enterprise* on *Star Trek* or the famous Robert A. Heinlein sentence, "the door dilated," in the first paragraph of *Beyond This Horizon* [1942]). In these classical versions of the future, humans are more or less the same and rely on their innate ingenuity, creativity, and new technological tools to solve the various problems posed by whatever new horizon is being explored. In this kind of speculative story, humanity is presented amongst new technologies but is for the most part unchanged. Transhumanism, instead, articulates a new kind of futurity, one in which humans are rendered into a kind of posthumanity[7] through technologies that fundamentally alter basic elements of the human condition: lifespan, morphology, cognition. Rather than evolving our future's mise-en-scène, transhumanism represents a new vision of the future in which we are ourselves being evolved by the futuristic setting we have already created for ourselves.

While transhumanism is often dismissed in academic circles as a retrogressive assertion of Cartesian humanism, a techno-secular reimagining of Christian fundamentalist salvation history, and a celebration of the most brutal forms of capitalist excess in the present, I argue in this book that transhumanism is also what utopian thought *might* look like in the age of network culture, big data, and the quantified self: a utopian rhetoric for the information age. As cutting-edge post-Marxist theorists lament the failure of state-centric utopia in the face of neoliberal capitalist expansion—most notably in Franco Berardi's *The Soul at Work* and Fredric Jameson's *Archaeologies of the Future*—transhumanism helpfully articulates a rhetoric of utopia that uses the human body and the human soul, which Berardi claims is colonized by capital, as the material for imagining futures that are not mere re-presentations of the present. Although there are flaws in transhumanist argumentation—its alien rationality and its problematic defense of a myriad of concepts abandoned by avant-garde philosophy being just two examples—it is the only discourse today actively imagining a radical future as radically alien as communism's idea of a classless society was in the late nineteenth century.

To recover this core of utopian argumentation, I argue for considering transhumanism from a much longer perspective than the movement itself chooses to acknowledge. Specifically, transhumanism is part of a utopian rhetoric of technology I call evolutionary futurism. Evolutionary futurism, of which transhumanism is one heir in the present, rhetorically situates technology as exerting mutational, evolutionary pressures on the human organism. However, as I argue here and expand on throughout the book, the longer history of evolutionary futurist rhetoric focuses more explicitly on the ways of being—new philosophies, new social orders, new affordances—unlocked by an evolutionary overcoming of the category of the human. Many of transhumanism's most alien features, especially its problematic entwining with consumer capitalism, emerge from a fetishization of the material pathways toward an evolutionary future at the expense of the sweeping philosophical changes mandated by these very imagined, radical futures.

To connect the terms I am working with—transhumanism, evolutionary futurism, and utopia—I move through each in its turn in this Introduction, building up an argument that connects contemporary definitions of transhumanism to my concept of evolutionary futurism as a long-view, unifying radical thought in the nineteenth and twentieth centuries with calls for new, post-Marxist utopian praxis in the twenty-first. This Introduction, thus, constitutes a sketch of the theoretical terms in operation throughout the rest of the book and also is a defense of transhumanism as an avenue of intellectual inquiry. Rather than merely dismissing the movement out of hand for not conforming to state-of-the-art theories about the end of the human and the limits of reason, this book instead considers transhumanism as a font of utopian rhetoric that, however flawed, must be taken seriously if we are to understand utopia today.

Although still a relatively small and fringe movement in terms of core believers, the transhuman perspective on culture filters into our lives as an ideology of technology. As an example, though not explicitly transhuman in the same way as something like mind uploading or radical life extension, the quantified self movement, in which participants treat their bodies as data sets and attempt to optimize this data, responds to the transhuman perspective that the human is nothing more than an information pattern that happens to be currently instantiated in fleshy form. Moreover, recent documentaries such as *The Immortalists* (2013) and *Transcendent Man* (2011), about prominent transhumanists Aubrey de Grey and Raymond Kurzweil, respectively, have further popularized the basic beliefs of the movement, especially as both films have been prominently featured on the extremely popular streaming service provided by Netflix.

Against this backdrop, More's assertion that technologies will and are radically altering "the nature and potentials of humans" in our current age articulates an inherently evolutionary version of futurism. In this new version of the future, humans are subject to intense evolutionary pressures exerted by the external technologies we created to ease our lives in the first place. In many ways, transhumanism, as it first began to articulate itself in works such as Robert Ettinger's *Man into Superman* (1972)[8] and FM-2030's *UpWingers* (1973), stands in answer to the challenges and perils documented by Alvin Toffler in *Future Shock* (1970), an extremely influential bestseller addressing "the shattering stress and disorientation that we induce in individuals by subjecting them to too much change in too short a time."[9] Toffler argues that "the acceleration of change in our time is, itself, an elemental force" that leads to a "psycho-biological condition" that "can be described in medical and psychiatric terms."[10] He suggests that we suffer, and will suffer more in the future, from a "disease of change," that we "are doomed to a massive adaptational breakdown."[11] Toffler's book goes on to document the torments and possible solutions to what he saw as a too-rapid pace of change.

Perhaps, then, responding to Toffler's concept of a "disease of change" explains some of the oppositional tone of the earliest transhuman works (most of which were published in the ten years following *Future Shock*) and the rational structure More uses throughout his philosophical oeuvre. Where Toffler situates future shock as a kind of hysteria, an outburst of irrational unconscious energy, the often mollifying tone of transhuman philosophy suggests that the radical future is nothing to be worried about. Writing in *Man into Superman*, Ettinger is dumbfounded by the response of potentially future-shocked people who refuse to believe in his transhuman vision:

> Today there are vast segments of the world population that will not concede it is better to be rich than poor, better to be bright than dull, better to be strong than weak, better to be free than regimented, or even that it is better to live than to die.[12]

This strong moral imperative swirls around transhuman discourse into the present. "In the face of the evolutionary imperative posed by digital technology," transhumanism seems to ask, "why would you choose to die?" For many transhumanists, the availability of these technologies suggests an imperative to use them to remake ourselves into something more than merely human.

Talking about radical futurity as though it were the most obvious thing is, as one might suspect, not without its critics. Prominent transhumanist Hans Moravec was famously eviscerated in the opening of N. Katherine Hayles's influential *How We Became Posthuman* (1999) as her first example of the transhuman tendency to declare human embodiment passé at the moment in which postmodern feminists were specifically contesting patriarchy through questions of embodiment. Similarly, Cary Wolfe dismisses the movement in *What Is Posthumanism?* (2009) by saying that transhuman arguments derive "directly from ideals of human perfectibility, rationality, and agency inherited from Renaissance humanism and the Enlightenment."[13] For Wolfe, statements such as More's definition ("the continuation and acceleration of the evolution of intelligent life beyond its currently human form and human limitations") stand in opposition to the basic aims of a truly posthuman philosophy. For Wolfe, "'the human' is achieved by escaping or repressing not just its animal origins in nature, the biological, and the evolutionary, but more generally by transcending the bonds of materiality and embodiment altogether."[14] Wolfe's "sense of posthumanism is the *opposite* of transhumanism" because transhumanism is merely "an *intensification* of humanism."[15]

As I argue below, Wolfe's claims about transhumanism are not entirely representative. Moreover, transhumanism is more complicated than just "an intensification of humanism," as we have already seen in Max More's argument that transhumanism is also a radical refiguring of what it is possible to do as a human. The claims of the movement, though, have a variety of dangerous and dubious rhetorical associations and, to better understand the potential affordances of transhuman thought, we need to first consider the major criticism of transhumanism: its complicated and often contradictory relationship to Christian salvation history.

Transhumanism as a Religion of Technology?

One feature of transhumanism, the Singularity (which I discuss in chapter 3), has been called by SF writer Charles Stross "The Rapture of the Nerds," connecting the transhuman faith in technology's radical evolutionary potential to Christian theories of eschatology. Both these futurisms await the arrival of a savior; where Christian eschatology awaits the return of God, transhumanism awaits the arrival of a host of radical posthuman-making technologies. While several academics in the field of religious studies have recently published criticisms of transhumanism and religion,[16] Stross's pithy statement remains the most viral figure of the strong methodological similarities between transhumanism and religious salvation history. Despite the vehement secularism of most in transhumanism (More declares religion "entropic" and focuses on the "loss of information" in his 1996 definitional essay), much of the criticism has focused on the millenarian, eschatological features of transhumanism.[17] Transhuman argumentation *does*, in an overly reductive mode, boil down to a faith that technology in the future will arrive and save us, not unlike the Second Coming in Christianity. More's argument about transhumanism, violently opposed to religion as it is, problematizes this similarity by performing something of a bait and switch by suggesting that transhumanism will "replace religions with other types of meaning-fostering systems," rather than getting past the outmoded need for faith itself—not unlike the process by which alcoholics transfer their addictive behavior to support groups such as Alcoholics Anonymous instead of curing their addiction itself. This idea, merely replacing one system of faith with another, is one of the central criticisms of the problematic relationship between transhumanism and religion.

Further intensifying this critique of transhumanism, David F. Noble argues in *The Religion of Technology* (1997) that the idea of religious salvation through technology dates back to at least the Middle Ages. As Noble suggests, in that period the "most material and humble of human activities become increasingly invested with spiritual significance and a transcendent meaning."[18] Technological progress, instead of divine contemplation, became increasingly associated with the return of a lost human perfection.[19] Noble's argument traces how, over the course of Western history, even as the explicit connection between religious salvation and technological progress became opaque, the belief in the transcendent potential of new technologies clings to modernity and the project of technological progress. In the latter half of his book, he traces these themes through nuclear weapons, space travel, artificial intelligence, and genetic engineering (themselves key transhuman technologies) and shows how these technologies seem to their creators to fulfill key aspects of a divine destiny for humanity.

Given Noble's thesis, what is novel, then, about transhumanism? Though *The Religion of Technology* deals with many of the central technologies associated with transhumanism and though the book is often cited as being relevant to transhumanism, Noble does not discuss transhumanism explicitly in the volume. So while transhumanism may be a part of Noble's religion of technology, the movement also suggests two changes to Noble's thesis: One, transhumanism argues for an internalization of technology into the body. Two, transhumanism specifically explores what bioethicist Nicholas Agar labels "radical human enhancement," the increase of human potential well beyond what is possible to even imagine the human body performing. Unpacking both these distinctions will further clarify the newness of transhumanism with regard to thinking of it as a religion of technology.

Transhumanism represents a cultural shift in which the technologies changing the horizon of our lives have a significantly more intimate relationship to our bodies. As the transhuman technologies we saw in More's list ("neuroscience and neuropharmacology, life extension, nanotechnology, artificial ultraintelligence, and space habitation") become increasingly a part of our lives, they also become part of our bodies. This internalization of technologies suggests that transhumanism mutates Noble's understanding of the religion of technology in profound ways. Most of the technologies discussed in Noble's book are external to the human body and, importantly, work by analogy in his argument. He cites moments in which the rhetoric of scientists uses metaphors for religious phenomena in which the saints fly up to heaven *like* a rocket ship and God ends the world in Revelations *like* an atomic bomb. For Noble, these metaphoric relationships satisfy the original religious impulse. Reading the more radical claims of transhuman thinkers, however, instead of providing technological change to metaphorically emulate the power of the gods, transhumanism's promise is much more radical transcendence at the unit of the organism itself. While artificial intelligence, in Noble's reading of its history, might provide an analog for the immortality of gods or souls in various religions, the transhuman technology of radical life extension promises to make the user literally immortal, without the need for a metaphoric compromise.[20]

Where technologies before the emergence of an evolutionary version of the future fulfilled, on some unconscious level, a desire for godlike transcendence, the transhuman suite of technologies More lists in his definition are designed to be incorporated into the body with the evolutionary upshot being human transmutation into a literal godhood. For instance, biologist and life-extension researcher Aubrey de Grey suggests in a 2005 interview that "it's reasonable to suppose that one could oscillate between being biologically 20 and biologically 25 indefinitely."[21] De Grey, head of the Strategies for Engineered Negligible Senescence (SENS) project at Cambridge University, goes on to suggest that technology enabling humans to live for as long as a thousand years may be less than twenty-five years away from general availability. Further, de Grey suggests that aging and death, rather than the natural ordering principles for a human being's life, are the product of a flawed biology that can be corrected through the application of technologies of radical life extension. De Grey's life's work has been attempting to reverse engineer the processes of aging with an eye toward hacking the human body to remove them. Successful outcomes from this hacking literally, rather than metaphorically, restore the human body to a state of Edenic purity, in which the flesh would never decay.

By literalizing the metaphoric promise of transcendence inherent in technological striving, transhumanism transitions humanity from tool user to tool imbiber. The body that thus consumes these transhuman technologies becomes a technological body on orders of magnitude previously unimaginable in human history. This technological body not only means an end to the reasonably stable understanding of the biological human but also mandates a rethinking of the core philosophical values of humanism. So, in this way, transhumanism seeks doubly to overturn notions of bodily integrity to create these posthuman beings while simultaneously creating a speculative ethics to govern the actions of these hypothetical posthuman beings.

Related to this idea of a speculative ethics for hypothetical beings, the second transhuman addition to Noble's theory of the religion of technology lies in how transhumanism, according to bioethicist Nicholas Agar, intensifies the general technological program of human enhancement into what he calls "radical enhancement." In contrast to acceptable (to Agar) human enhancements such as antibiotics and automobiles, Agar argues that "radical enhancement involves improving significant human attributes and abilities to levels that greatly exceed what is currently possible for human beings."[22] In *Truly Human Enhancement* (2013), Agar illustrates the difference between beneficial and radical enhancement technologies by discussing "veridical engagement," what "we can accurately imagine ourselves doing as" we watch others perform feats of skill and strength.[23] He discusses Usain Bolt, record-breaking track star, and the Flash, fast-running hero of DC comics, to explain veridical enjoyment.[24] When Bolt breaks world records on the track, we watch and identify with his struggle; although his feats amaze us, we can imagine *a* body performing them, even if it is not *our* body. With the Flash, however, his speed (which often exceeds the capacity of the human eye to observe, even allowing him to time travel) is so great that we cannot identify with his actions. The Flash represents something that is inhuman, while Bolt's speed is recognizably human. For Agar, the difference between beneficial human enhancement and radical human enhancement lies in this ability to recognize and function within current limits. In Agar's bioethical framework, all of us running as fast as Usain Bolt thanks to robotic legs would be on the border of acceptable enhancement, while running as fast as the Flash would not because we would no longer be able to veridically engage as human.

As we saw in Ettinger's perplexity at those who choose to die, much of the transhuman community is deeply invested in the internal, radical changes suggested by the types of technology discussed by More. Many in the movement find Agar's bioethical arguments outmoded and akin to nostalgia for a form of existence in which humanity is enslaved to death and aging. With this internalization and this radicalization, the idea that we (whether as individuals, nation-states, or species) control our destiny becomes more and more likely. Transhumanism envisions the world as one in which Nature is no longer driving our species' evolutionary bus.

Max More, writing in 1999 in "A Letter to Mother Nature," captures this idea best in his apostrophe to the planet. He writes,

> What you have made us is glorious, yet deeply flawed. You seem to have lost interest in our further evolution some 100,000 years ago. Or perhaps you have been biding your time, waiting for us to take the next step ourselves. Either way, we have reached our childhood's end.
>
> We have decided that it is time to amend the human constitution.[25]

The amendments he goes on to propose include standard transhuman *topoi* such as no longer tolerating "the tyranny of aging and death," supplementing "the neocortex with a 'metabrain,'" and asserting that "we will no longer be slaves to our genes."[26] In all these cases, we, More argues, must evolve ourselves in the face of a mother who has "lost interest" in her children.

More's use of "childhood's end" in talking to Nature about transhumanism is particularly telling. The phrase is also the title of an Arthur C. Clarke novel from 1953 in which an alien race called the Overlords arrive on Earth to steward our species' transition into a new age of evolutionary existence as beings of pure energy. In the novel, Clarke articulates a number of the major themes that become associated with transhumanism and give transhumanism one of its most potent images. From the star child at the end of *2001: A Space Odyssey* (1968) to the titular hero of *Lucy* (2014), the trope of an evolutionary explosion and the emergence of humanity's new evolutionary destiny is constantly imagined and presented as a maturation from infancy to full adulthood. Like the child that must inevitably leave the nest, More argues in his letter, humanity has come into its adulthood, ready to make decisions about its own destiny.

This sense of destiny is, finally, what signals transhumanism as a mutation in Noble's religion of technology. Where Noble suggests a desire to satisfy God's plan for humanity through technological enhancement, transhumanism instead appears to argue that we will make ourselves into gods. This internalization of control is one of transhumanism's most interesting features: as More also makes clear, even a blind external force, such as evolution, is too little control in the hands of humanity.

Evolutionary Futurism

Up until this point, we have been discussing transhumanism in a number of terms: as a movement, as a philosophy, even as a new mutation in the religion of technology that has driven Western progress narratives since the Middle Ages. Transhumanism, we can conclude, has a complicated definitional history. Max More insists that transhumanism is a philosophy, as we have seen. However, many transhumanists have defined the movement as a mode of living that is at home in the future.

But what would it mean to be "at home in the future"? We might answer this question by attuning ourselves to the medium in which transhumanists project their futuristic habitation, namely language itself. While transhumanists argue that their vision of the future will come to pass, this future is, as all futures ultimately are, a linguistic mirage, a projection of a particular configuration of technoscientific knowledge and ideological hopes that coagulate into a coherent image of tomorrow. In other words, we can productively define transhumanism by taking recourse to the rhetoric it uses to create this futuristic vision. Thinking about linguistic projection, rhetorician John Poulakos, in arguing for the rehabilitation of the ancient rhetoric teachers known as the Sophists, defines rhetoric as "the art which seeks to capture in opportune moments that which is appropriate and attempts to suggest that which is possible."[27] He clarifies the Sophists' use of the possible (*to dynaton*) as a strategy that "affirms in man the desire to be at another place or at another time and takes him away from the world of actuality and transports him in that of potentiality."[28] Thus, as Poulakos argues, rhetoric for the Sophists is not the act of changing minds but of presenting possible, desirous futures.

Having said that, in this book I do not consider transhumanists *as* rhetoricians (and certainly not as Sophists). Instead, by focusing on their use of *to dynaton* to project the desirability of *a* future, I want to imagine transhumanism as a rhetorical mode, a means of creating and seducing through language about the future. By rhetorical mode, I mean, as Chaïm Perelman and Lucie Olbrechts-Tyteca define in *The New Rhetoric,* the "way in which we formulate our thought [that] brings out certain of its modalities, which modify the reality, the certainty, or the importance of the data."[29] In other words, a rhetorical mode is the way of mapping the general flux (raw data) of experience into a specific program for action. Perelman and Olbrechts-Tyteca's study of the "affective categories" of language and "the modalities of thought underlying variable grammatical forms" highlights how rhetorical choices—of words, of grammar, of tropes, of figures—shape the cultural reality of any utterance and how repeated patterns of these choices emerge as modalities of suasive communication.[30] To use Poulakos's terminology, a rhetorical mode, then, is the way of projecting a particular future through language.

To turn specifically to the rhetorical modality of transhumanism, Timothy Morton suggests that rhetorical modes are "affective-contemplative techniques for summoning the alien," which is an image of the function of language that speaks specifically to what transhumanists do when they talk about their version of the future.[31] Rhetorically contacting what Morton calls the "the strange stranger," transhumanist language offers a series of linguistic operations that project near-future evolutionary change *and* position technical artifacts as the vectors for producing this imminent overcoming. It is also a way of expressing the inevitability of this radical futurity and its desirability. As we have already seen, transhumanism offers a uniquely alien vision of the future that is simultaneously made less strange through a variety of rhetorical strategies.

In tracing this rhetorical mode, this project summons an even more alien version of transhumanism. While, as I have been arguing up to this point, transhumanism uses a rhetorical mode that sees evolutionary change being mediated by technological progress, the larger project of this book is to document how this rhetorical mode actually predates the mid-1970s organization of transhumanism into a coherent movement. Instead, I argue that this older rhetorical mode is more productively labeled as "evolutionary futurism"—a set of rhetorical strategies meant to depict a future in which our machines evolve us beyond our current human limitations—and that it organizes a wildly disparate swath of twentieth-century intellectual history, including avant-garde modernism, early science fiction, poststructural philosophy, and evolutionary synthesis in biology. In tracing this rhetorical mode through these various sites, I examine moments that may not specifically appear as "transhuman" in the orthodox sense. In doing so, I show the rich and varied rhetorical practices that contribute to the emergence of evolutionary futurism as a rhetorical mode and that help lay the foundation for transhumanism as an organized movement of evolutionary futurists in the present. Additionally, I route many of these explorations through philosophers, such as Friedrich Nietzsche and Gilles Deleuze, not specifically associated with transhuman philosophy but who have been connected with arguments about the limits of the human and the transcendence of these limits.[32] This method is not intended to dilute transhumanism as an intellectual movement; instead, I argue that evolutionary futurism shapes broad and fruitful discourses that we ignore when we reject transhumanism as a serious intellectual endeavor.

Methodologically, I enact a series of conversations between contemporary transhuman philosophy and this larger, older body of evolutionary futurist rhetoric. In doing so, I explore the more nuanced valences of this rhetorical mode outside the perceived flaws in orthodox transhumanism. Additionally, these conversations document other rhetorical choices that may not have been taken up into the core of transhuman argumentation. By focusing in this way on the wider universe of rhetorical possibility contained in evolutionary futurism, I show how transhumanism contains a potent rhetorical core that allows for the rethinking of utopia in the age of informatics, specifically by shifting our attention from the state to the body as the needed site of utopian investment. To further clarify this point and to suggest the broader method deployed throughout this book, in the remainder of this introduction I offer an example of a discourse (automata theory) that uses evolutionary futurist argumentation but is not commonly hailed as orthodox transhumanism, and then conclude by connecting my interest in evolutionary futurist rhetoric to a crisis in contemporary conceptions of utopia, arguing that this rhetoric is a rhetoric of utopia. In connecting evolutionary futurism to utopian thought, I show how, in tracing a broader intellectual history of transhuman argumentation, evolutionary futurism constitutes an emergent and contemporarily relevant rhetoric of the future, one that reinvests the utopian imaginary into the human body. In an age dominated by theories that hinge on the failure of the welfare state as a utopian experiment, this rhetorical shift (from the state to the body) offers a potential means to reanimate speculation about spaces beyond the current configuration of power.

Evolutionary Futurist Rhetoric in Action: Automata Theory

As an example of something not classically hailed as transhuman but involved in the rhetorical work of transhumanism, John von Neumann's paradigm of the body as information is key to the emergence of contemporary transhumanism. Von Neumann, a polymath and major figure in computer science whose work with the EDVAC team at the University of Pennsylvania led to the publication of *First Draft of a Report on the EDVAC* (1945) as well the formalization of the modern computer architecture that still bears his name and structures most computers in use today, was interested, toward the premature end of his life (von Neumann died at 53 from bone cancer, likely contracted from his presence at the first thermonuclear tests on Bikini Atoll in 1946), in the interconnections between biological and computational processes. While obviously not a transhumanist (he died well before the movement began in the 1970s), his work shares many similarities to, and, as I argue in this section, stands as an important example of, the rhetorical moves an evolutionary futurism can make.

Von Neumann's contribution to the movement is not entirely ignored in the transhuman literature, specifically as Raymond Kurzweil, in *The Singularity Is Near*, cites a conversation between Stanislaw Ulam and John von Neumann as being the first mention of a technological singularity.[33] Additionally, despite this limited contribution, I suggest von Neumann's late work is an important rhetorical precursor to the transhuman vision of the future of the body. Relatedly, considering von Neumann's theories of computation in this way focuses our attention on the mutational aspects of this rhetorical mode.

Late in his life, von Neumann turned his attention to biology and psychiatry in an attempt to further intensify his theories of computation. Describing "automata theory," von Neumann argued that all processes of sufficient complexity, biological or computational,

> can be viewed as made up of parts which to a certain extent are independent, elementary units. We may, therefore, to this extent, view as the first part of the problem the structure and functioning of such elementary units individually. The second part of the problem consists of understanding how these elements are organized into a whole, and how the functioning of the whole is expressed in terms of these elements.[34]

For von Neumann, this insight suggested the possibility of easily transferring biological processes to computational hardware, the foundational narrative for the so-called strong theory of artificial life.

Summarizing this work on automata, technology journalist Mark Ward suggests that "von Neumann saw that it was the manipulation of information that keeps an organism alive, allows it to beat back entropy, and that life is a process not a property."[35] These insights are key to what would become transhumanism, especially for extropians such as computer scientist Hans Moravec and technologist Raymond Kurzweil who both map a transhuman future in which humans will upload their consciousnesses into machines. A quote popularly attributed to von Neumann suggests that the upshot of this work on automata was a view that "life is a process which can be abstracted away from any particular medium."[36]

By understanding life as a process that can be serialized between media (presumably between biological wetware and computational hardware), the idea that the body *is* information entered cultural currency. This transformation and its development in first- and second-order cybernetics is traced through N. Katherine Hayles's *How We Became Posthuman* (1997), but is also important for the emergence of particular transhuman *topoi*. Building on von Neumann's work, for instance, in his 1987 manifesto of artificial life, Christopher Langton suggests a linguistic move from "machine," as von Neumann used it, to "algorithm," as Langton defined it, "the logic underlying the dynamics of an automaton, regardless of the details of its material construction."[37] At this point, we see an outline of the completion of the informational body of transhumanism: if life is an algorithm, then we can tinker with or hack that algorithm to better optimize its outputs (us) along various ends.

By understanding thought as a mechanical process, and specifically as a process of simple, cellular automata working in concert to produce complex phenomena, von Neumann not only opened up this algorithmic understanding of cognition, he also suggested a specifically evolutionary futurist take on computational theory. As his editor Arthur W. Burks describes in the introduction to *Theory of Self-Reproducing Automata*, working to build ever faster computers, von Neumann sought to avoid the extreme unreliability of the components, such as vacuum tubes, "not by making them more reliable, but by organizing them so that the reliability of the whole computer is greater than the reliability of its parts."[38] He was specifically interested in this, reports Burks, because he wanted to produce artificially intelligent machines in the late 1950s, using the unreliable components at hand. After von Neumann's death, his radical work was abandoned in favor of building more reliable and faster components with an eye toward producing intelligent machines some time in the future. For von Neumann, the rudimentary computers such as ENIAC suggested a mutated view of the human, and this new paradigm of human cognition further suggested a near-future evolutionary leap in technologically mediated humanity.

In any case, von Neumann's work on automata theory, algorithmic consciousness, and vacuum tube AI provides an example of early computation immediately feeding back into mutations within the human organism. Our technologies external to the body come to color our understandings of ourselves in a way that focuses on near-future evolutionary change. Beyond the ideas of human perfectibility associated with Cartesian humanism, automata theory's use of a computational paradigm to describe the human body suggests a means for attaining this perfectibility, as well as showing the move from external tool development to internal tool imbibing.

Transhumanism and the Desire Called Utopia

Despite a serious commitment to imagining a new world and building a discourse to shape a better future, transhumanists are quick to assert that they are not utopians. There are a number of reasons suggested in the literature for this perspective; however, the issue of transhumanism as or as not a utopian movement came to a head in the 2011 edited collection *H+/-: Transhumanism and Its Critics,* in which a number of academics working on transhumanism wrote position papers to which prominent transhuman thinkers wrote rebuttals. A number of the academic position papers used the word *utopia* to describe transhumanism, and the responses from the transhumanists to this issue are highly suggestive of a terminological disjunction between academic and transhuman thinkers. In "From Mind Loading to Mind Cloning," Martine Rothblatt suggests that "the transhumanists are no more utopian or naïve than were the sociotechnological pioneers of the nineteenth century who believed in and fought for universal education, railroads, and public health."[39] While Rothblatt's ethos throughout the entire essay is similar to Robert Ettinger's earlier bafflement about those who would choose to die, the specific juxtaposition—"utopian or naïve"—is most emphatic on the issue of "utopia" in transhuman thought.

Stemming from Thomas More's definitional 1516 book, utopia has been, at the level of the word itself (literally "no place" in Latin), connected to notions of daydream, fancy, or idle speculation. Transhumanism seeks to avoid the idea of the lazy speculator, imagining future wonders: Rothblatt's juxtaposition suggests that utopian thought is naivety itself and that transhumanism is, instead, a philosophy of action. Moreover, Rothblatt focuses on the reality of the possibility that transhuman rhetors seek to impart: "this is not speculation," her dismissal of the utopian label suggests, "what we talk about *will* happen." Utopia—in this more popular understanding (tied to a contemporary instantiation in science fiction but indebted to a much longer history)—is an unreal, unrealizable imaginary future. Transhumanists, long disparaged in academia, are particularly sensitive to rhetorical moves that distance the movement from a practical, serious, and, most important, realistic future.

Also suggestive of the transhuman attitude toward utopianism is Max More's response to the "utopian" label in *H+/-*. He suggests that "True Transhumanism" (his essay's title) does "not see utopia or perfection as even a *goal.*" He continues,

> The Idol of Paradise and the idea of a Platonically perfect, static utopia, is so antithetical to true transhumanism that I coined the term *extropia* to label a conceptual alternative. Transhumanists seek neither utopia nor dystopia. They seek perpetual progress—a never-ending movement toward the ever-distant goal of extropia.[40]

More makes two important assumptions here: First, he suggests that utopia implies the creation of a static state, in which no more progress is possible. Second, he suggests that, instead, transhumanism values a continual progress narrative in which a perfect state is never possible to reach.[41]

Despite More's rejection, I want to use his idea of this "ever-distant goal of extropia" to claim that while transhumanism may not be utopian, it is definitely Utopian. My use of the lowercase and uppercase to draw a distinction may seem too facile, but I do so to signal a potent distinction in the concept of Utopia itself. Written with the lowercase, "utopia" connects to the so-called utopian socialists that Marx and Engels use as one group of conceptual enemies in the emergence of Marxist theory, and generally to the idea of naive speculation rejected by Rothblatt.[42] "Utopia" written in uppercase is the collection of processes for imagining spaces beyond capitalism that feed into the practices of political resistance in the present, especially in the works of Fredric Jameson, where he wants us "to understand Utopianism . . . as a whole distinct process" rather than the creation of a specific place or scheme.[43]

In Jameson's work on Marxism in the age of the postmodern, "Utopia" comes to mean a kind of methodological imagination. Primarily detailed in *Archaeologies of the Future* (2005), the discussion of Utopia throughout Jameson's career has foregrounded it as a "mechanism" for the speculative production of new modes of being.[44] As he asks in "Utopia as Replication," "we ordinarily think of Utopia as a place, or if you like a non-place that looks like a place. How can a place be a method?"[45] In answer to this, Jameson suggests that in culture there exists an "obligation for Utopia to remain an unrealizable fantasy."[46] Jameson's understanding of Utopia as a method and a fantasy—he would label it "the desire called Utopia," using the Lyotardian formula so prevalent in his more recent work—shows his intense debt to the cybernetic psychoanalysis of Jacques Lacan, for whom the object of fantasy always recedes from our graspings after it. For Jameson, Utopia is a method of imagining schemes, spaces, objects, whole futures that, through the power of fantasy and desire, inspires political change in the real world but never actually arrives from the fantastic into the real.

This idea of an endlessly receding horizon is also what More means by extropia. Thus, the progressive creation of new social forms is what makes transhumanism Utopian, and specifically, as I discuss below, a form of Utopia uniquely relevant in an age when paths not moving toward the cybernetic brutality of multinational capital appear to be exhausted. In both Utopia and extropia, we find a process by which humanity can constantly exceed its current limits and imagine future forms for itself. Jameson is also careful, like More, to build a method that sees an interchange between and finds inspiration in the everyday, suggesting that any "Utopian enclave" (Jameson's term for a site of speculative, Utopian investment, such as money in the Early Modern Utopias or the Internet in the early 1990s) eventually "becomes 'only that,' descending from a transcendental ideal into a contingent set of empirical arrangements."[47] This move between the fantasy (the ever-exceeding goal More labels "extropia") and the quotidian is important for Jameson's thought and his analysis of culture, where even degraded cultural forms such as advertisements or the inartistic films he often discusses[48] are able to articulate "the oldest Utopian longings of humankind."[49]

However, this connection between transhumanism and Jameson is not merely terminological, nor even just methodological. We can construct a transhuman Jameson based on the language of mutation and cognitive expansion prevalent throughout his work. A transhuman Jameson's investment in the Utopia as a place that is also a method spills over into a belief in evolutionary futurism as the only possible avenue of revolutionary politics in the present moment of globalized capital. In *Archaeologies of the Future,* he suggests that his vision of Utopian speculation "is probably on the side of the imagining of the post-human."[50] Jameson continues by adding that Utopian speculation may be on the side "even of the angelic" as well as the posthuman. By using two models of human perfectibility, I argue, Jameson uses the term here in the way the transhumanists use it: an imagined placeholder for what comes out the other end of their transformative processes. This contrasts sharply with notions of the word in, say, Cary Wolfe's *What Is Posthumanism?* which allies the term with a post-Cartesian openness to the Other. This association between

Utopia and a transhuman posthuman further intensifies the reality of a transhuman Jameson. The pattern of usage of "post-human" in *Archaeologies*, mostly in the chapters "Journey into Fear" and "Synthesis, Irony, Neutralization, and the Moment of Truth," situates that strand of inquiry within the longer tradition of Utopia, suggesting that to imagine beyond the human is, for Jameson, specifically Utopian.[51] In encapsulating the argument of the book, before moving into its equally enigmatic conclusion, Jameson suggests,

> We have come laboriously to the conclusion that all ostensible Utopian content was ideological, and that the proper function of its themes lay in critical negativity . . . In addition we have been plagued by the perpetual reversion of difference and otherness into the same, and the discovery that our most energetic imaginative leaps into radical alternatives were little more than projections of our own social moment and history or subjective situation: the post-human thereby seeming more distant and impossible than ever![52]

In concluding this summary of his argument about the history of Utopia as he does, Jameson seems to suggest that the move beyond the human, the creation of posthuman existence, is the end point of Utopia, especially in the present.

In addition to this discussion of the posthuman as goal for Utopia, Jameson's most direct engagement with evolutionary futurism occurs in the famously puzzling reading of the Bonaventure Hotel in Los Angeles in *Postmodernism.* Lost amid the challenging architecture of John Portman's dizzying and maze-like lobby, Jameson declares "the newer architecture therefore . . . stands as something like an imperative to grow new organs, to expand our sensorium and our body to some new, yet unimaginable, perhaps ultimately impossible, dimensions."[53] From merely speculating on the creation of a "post-human" through Utopian thought, here Jameson begins to imagine the creation of these new beings, with new organs and, most important, new cognitive capacities. Cognitive mapping is Jameson's specific response to the Bonaventure's call to grow new organs, to evolve beyond our present form. For Jameson, this cognitive mapping is the Utopian response to the "perceptual barrage of immediacy" and the related "fragmented and schizophrenic decentering and dispersion of" the human that makes up life in the postmodern.[54] Like transhumanists arguing that our technologies of global communication exceed our philosophies for understanding them, Jameson suggests that our narratives of self and state, as well as our ability to grow either, are fundamentally broken amid this perceptual barrage that we lack the expanded sensorium to process. The challenge of the postmodern—indeed the challenge of the Bonaventure or the dizzying media landscapes of sculptor Nam June Paik or many of the other texts Jameson famously reads in *Postmodernism*—"often takes the form of an impossible imperative to achieve that new mutation in what can perhaps no longer be called consciousness."[55] This "new mutation," potentially growing "new organs," is, I argue, the core of articulating Jameson as a thinker committed to an evolutionary futurist mode of argumentation. Moreover, in connecting these two quotes, we see Jameson articulating the properly transhuman argument that changes in morphology yield to changes in cognition, as Max More articulates in his definition of the movement.

Jeffrey Nealon, in his heavily Jamesonian *Post-Postmodernism*, glosses the "new mutation" quote in terms that map "consciousness" to something like class consciousness in Marxist theory or a consciousness of aesthetic forms.[56] Similarly, Wendy Hui Kyong Chun, in *Programmed Visions*, glosses the "new organs" quote as being about our need to "grasp our relation to totality—to make sense of the disconnect between, and possibly to reconnect, the real and the true."[57] I read both these comments as taking Jameson's call for mutation and new growth as metaphors for the need to rewire our thought for a new age of the political. But I ask: What if Jameson means this literally? Literal new organs, literal mutations in cognition. If we take Jameson at his word, instead of interpreting these calls metaphorically, we start to see the emergence of a properly transhuman Jameson, one who suggests that we must now evolve in response to our media ecology. In constructing this transhuman Jameson, we see that transhumanism is a Utopian system for imagining the new organs we might grow if we are to survive in the dizzying media landscape we have constructed for ourselves but do not yet understand.

Both Chun and Nealon articulate the desire behind this transhuman Jameson's call to expand our sensorium: there is a problem mapping our limited, human cognition into the global flows of information that make up postmodern life. This lack emerges as a crisis in the Utopian imaginary. In Nealon's reading, Jameson's Utopian project of imagining a "new organ" in the body and a "new mutation" in cognition was provoked by mutations in the exploitation of the human by capital, exploitative forms of power made possible by the kinds of technologies that inspire transhumanism in the first place. By looking back at and tracing forward Jameson's 1991 argument, Nealon, following Michael Hardt and Antonio Negri in *Empire*, suggests the postmodern was the emergence of a move within capitalism from an "extensive" economy, in which new markets are sought, to an "intensive" one, in which old markets are valued and exploited in new ways: "now [capital] turns inward toward intensification of existing biopolitical resources. The final product, in the end, is you and me."[58] In this era of capitalism, cognition is the chief site of labor, as Italian autonomist Franco Berardi has highlighted: "work is performed according to the same physical patterns: we all sit in front of a screen and move our fingers across a keyboard. We type."[59] As Berardi, Negri, and the other Italian autonomist Marxists have argued, the configuration of the economy as a cybernetic information system, at the core of this neoliberal economic revolution, has made the mind, its maintenance, and its discipline as important today as the discipline of the body was to modes of capital organized around industrial production.

Domesticated Evolutionary Futurism: Commodification of the Self versus Radical Thought in the Present

Transhumanism has had a problematic relationship with this economics of the soul. Specifically, this difficult association is related to a crisis in evolutionary futurist ethos. On the one hand, as we shall see in chapter 1, the more mystical versions of evolutionary futurism are often dismissed by mainstream transhumanists as lacking in rigor and seriousness. On the other, the much more rational ethos of someone like Robert Ettinger risks being rejected for attempting to be serious about something that, to many people, seems far-fetched. In both cases, either a radically mystical or a radically rational approach to evolutionary futurity comes off as rhetorically unseemly to many audiences. In attempting to solve this problem of ethos, contemporary transhumanists often subtract many of the more radical claims made in the long history of evolutionary futurist rhetoric to create what we might call a domesticated strain of evolutionary futurism. This domestic evolutionary futurism suggests worrying linkages to the neoliberal construction of subjectivity as commodity and to the reduced capacity for basic empathy seemingly integral to contemporary global capital (as when, for instance, Silicon Valley insiders suggest that food stamps might effectively be replaced with boxes of the engineered nutritional supplement Soylent, a product designed specifically to help users optimize their flawed bodies). Further, these moves to domesticate evolutionary futurism risk reducing the alien philosophy proposed by Max

More and other thinkers working in the evolutionary futurist mode to a kind of defanged technological solutionism. Separating this tendency to march in lockstep with neoliberal productions of the self from the Utopian core of evolutionary futurist rhetoric is a potent vector for reigniting radical thought in the present.

Berardi's work, especially in recent years, has focused on this machining of the unconscious as a tool for neoliberal subject-making, and, generally, we might extend his thinking along evolutionary futurist lines to suggest that, if the mind is at stake in neoliberal capitalism, a discourse designed to expand the mind into new and radical dimensions would be a potent antidote to this cybernetic machining. Unfortunately, contemporary transhumanism only partially provides such a curative. As the remainder of this book unpacks, there are many moments in the history of evolutionary futurism that anticipate a Utopian cognitive evolution. However, equally, the increasing trend toward domesticating the contemporary transhuman movement—stripping the more radical philosophical insights of someone like Max More—and focusing on a kind of technological solutionism risks aligning transhumanism with the neoliberal machining of the unconscious into a field of value production. Nealon's analysis of neoliberalism, in which ultimately the product produced "is you and me," highlights the commodification of the private and the use of branding as a means of packaging the self as both a product for sale and as chief consumer. In this way, Nealon highlights the centrality of lifestyle branding to contemporary economics.

The move to domesticate transhumanism, as a strategy to solve the movement's alien ethos, is not immune to creating a transhuman lifestyle brand. In FM-2030's *Are You a Transhuman?* (1989), the rhetoric of the movement shapes transhumanism as a kind of postindustrial lifestyle brand in which buying the right high-tech gadgets will make consumers into "the earliest manifestations of new evolutionary beings."[60] To help readers answer its titular question, FM-2030 also provides more than two hundred pages of *Cosmo*-style quizzes—"How Fluid Are You?" "How High Tech Is Your Attention Span?" "What Is Your Level of Humanity?"—to evaluate how close you are to "aligning and accelerating your rate of personal growth."[61] In doing so, FM-2030 accelerates the rational futurity of Ettinger and, instead of asking readers to be unafraid, asks them to be at home with a radical future. Buying products *becomes* radical evolution. In 2010, Raymond Kurzweil and the medical doctor Terry Grossman published *Fantastic Voyage: Live Long Enough to Live Forever*, a kind of self-help guide for readers to learn how to extend their lives long enough to capture the radical life extension technologies that were sure to arrive in the near future. Related to this publication, Kurzweil and Grossman also launched their company, Ray and Terry's Longevity Products, to market a variety of supplements and health foods—including Green Tea Extract and Melatonin—to help their readers attain this future immortality. However, as their website also makes clear, they are not merely another company selling herbal supplements. Their "About Us" page reads like a catalog of evolutionary futurist topics, exhorting customers to "reprogram your outdated system" and "take control of your own health status."[62] In effect, the company creates transhumanism as a kind of lifestyle brand in the same way that Apple effectively articulates its products into a lifestyle narrative to be purchased and inhabited. While aligned with the general New Age focus on healthy and holistic living, the company explicitly connects its product line to a lifestyle organized around transhuman goals: longevity, control, reprogramming, and so on.

When I can purchase longevity products from Ray and Terry to turn myself into a transhuman, I am intensifying my brand identity as a transhuman consumer, playing into the logic of capitalist intensities at the core of the neoliberal reprogramming of the unconscious: I manufacture myself as a member of the transhuman demographic. On the other hand, however, the more extropian visions of transhumanists, such as the one outlined by Max More, imply the subversion of the very logic of human subjectivity that this form of capital requires for its operation. This subversion is the reason, I argue, that Jameson's postmodern understanding of Utopia resonates so strongly with transhumanism. Although Jameson is staunchly committed to an idea of Utopia as being about an overcoming of the state,[63] ideas about the human as the conceptual limit in our era are everywhere in his work. Additionally, the shift in focus from the body to the unconscious, as suggested by the Italian autonomists, changes the necessary location of Utopian investment, and evolutionary futurism provides the concepts for imagining new political and cultural futures for humanity in which the human and not the state must be radicalized.

Despite this Utopian possibility, many in the contemporary transhuman movement see a more limited scope for what they hope to accomplish. I recently had a conversation on Twitter with a self-identified transhumanist who declared that transhumanism to him was just "people interested in using technology to help people," which is a classic definition of technopositivism, the philosophical belief that technology will always be able to solve any problem humanity needs to overcome. With the emergence of a whole swath of lifestyle artifacts associated with transhumanism—from Ray and Terry's various pills to Google Glass and Soylent to the latest in smart-home swag—there is a risk of diminishing the profoundly disruptive narratives about the future told by evolutionary futurism and turning this Utopian rhetoric into just another disruptive, solutionist ideology, thereby dangerously aligning transhumanism with core concepts of neoliberal subjectivity. So, in an effort to be taken seriously, how willing is transhumanism to dilute the content of its core insights? Moreover, what damage is done to the broader scope of evolutionary futurism, as a discourse of a Utopian future for our present moment, if this association between the neoliberal subject and the transhuman is intensified?

By recovering the older, weirder aspects of evolutionary futurism, we can more clearly see the contributions transhumanism can offer to thinking about Utopia in the present. I wonder—and hence this book's investigation into the rhetorical mode of transhumanism—if the limits of a certain configuration of the human also represent a limit to this kind of power relationship. In other words, I wonder if Max More's definition of transhumanism—"the continuation and acceleration of the evolution of intelligent life beyond its currently human form and human limitations"—is not also a way out of our current geopolitical bind. The core insights of evolutionary futurism suggest a variety of novel solutions for thinking beyond the blockage in the Utopian imaginary created by the failure of the welfare state. Transhumanism, then, is a Utopian rhetoric for an age of informational bodies and neoliberal subjects. Rather than dismiss transhumanism as naive or overly religious or too heavily invested in neoliberal subjectivity, this book instead argues that the multiheaded, polyvocal constellation of ideas we can label as transhuman are a goldmine for radical thought in the present.

We must start digging.

1 AN INNER TRANSHUMANISM

Modernism and Cognitive Evolution

When explaining the topic of this book to colleagues and friends, inevitably the topic of Nietzsche emerges: "Isn't this just the *Übermensch*?" they inevitably ask. Given Nietzsche's proclivity for imagining an overcoming of limiting human factors and declarations such as this one from *Thus Spoke Zarathustra*, "I want to teach humans the meaning of their being, which is the overman, the lightning from the dark cloud 'human being,'" there appears to be strong affinity between evolutionary futurist rhetoric and Nietzsche's philosophy of human overcoming.[1] The truth of this relationship, as this chapter will unpack, is that it is complicated.

In this chapter, I trace early instances of evolutionary futurist tropes through European modernism at the dawn of the twentieth century—specifically in the mystical account of evolution authored by P. D. Ouspensky and in the feminist futurism of Mina Loy—to illustrate Nietzsche's influence on the early formation of the rhetorical mode I call evolutionary futurism. Despite its Nietzschean heritage, however, the contemporary transhumanism movement maintains a much more fraught relationship with Nietzsche's philosophy. In a deliberately controversial passage of his generally divisive "A History of Transhuman Thought," philosopher Nick Bostrom dismisses the surface similarity many people note between Nietzsche and transhumanism:

> What Nietzsche had in mind, however, was not technological transformation but rather a kind of soaring personal growth and cultural refinement in exceptional individuals (who he thought would have to overcome the life-sapping "slave-morality" of Christianity). Despite some surface-level similarities with the Nietzschean vision, transhumanism—with its Enlightenment roots, its emphasis on individual liberties, and its humanistic concern for the

> welfare of all humans (and other sentient beings)—probably has as much or more in common with Nietzsche's contemporary J. S. Mill, the English liberal thinker and utilitarian.[2]

Bostrom is not wrong to make these claims. While contemporary instantiations of the transhuman project are primarily connected with evolutions of the body—through acceleration, simulation, and augmentation of the body's various systems (including the brain)—the longer history of evolutionary futurism is strongly marked by a focus on spiritual and cognitive evolution.

This second strand, what I will be calling "inner transhumanism," is more closely attuned to Nietzsche's concept of the *Übermensch*. Moreover, a focus on modernism in this chapter shows how, despite divergences and mutations further forward in time, Nietzsche was hugely influential in first suggesting the possibility of breaking from the human and the idea of evolutionary futurism. However, Bostrom argues that the stories of Gilgamesh's quest for immortality and the fountain of youth are also transhuman precursors, further discrediting Nietzsche as an origin point in his version of transhuman history.[3] That said, Nietzsche was important for the modernists I discuss in this chapter because his vision of overcoming human limits is posited in terms of a radical break from the very idea of the human. In the hands of theosophists, futurists, and similar European avant-gardes, Nietzsche's idea of a break from the human becomes coupled with modernist *topoi* that mark the cognitive pressures of industrialization, globality, and urbanization as forces driving us beyond the human. Specifically, P. D. Ouspensky and Mina Loy are the first among modernist thinkers to connect Nietzsche's philosophy to what are recognizable as contemporary transhuman arguments. Their rhetoric is strikingly similar to more current examples of evolutionary futurism, but, as I argue, they both focus on a kind of hybridization of spiritual/cognitive enhancement coupled to a technological reconfiguration of evolution from biological to machinic, combining both the more common outer transhumanism *and* a hugely important *inner* transhumanism.

Nietzsche's Transhumanism?

The quotation above from Bostrom's "A History of Transhumanist Thought" set off a specifically revealing exchange in the pages of *The Journal of Evolution & Technology* (*JET*), the premier peer-reviewed journal of contemporary transhumanism. Following the statement that transhumanism owes more to John Stewart Mill than to Friedrich Nietzsche, Stefan Lorenz Sorgner, a philosopher whose work grapples with Nietzsche's posthuman thought, published "Nietzsche, the Overhuman, and Transhumanism" in response. For Sorgner, Nietzsche's understanding of evolution beyond the human has much in common with transhumanism. He writes, contra Bostrom's insistence,

> Nietzsche does not exclude the possibility that technological means bring about the evolutionary step. His comments concerning the conditions for the evolutionary step toward the overhuman are rather vague in general, but in this respect his attitude is similar to that of transhumanists. However, he thinks that the scientific spirit will govern the forthcoming millennia and that this spirit will bring about the end of the domination of dualist concepts of God and metaphysics, and the beginning of a wider plausibility for his way of thinking.[4]

Sorgner's thinking accords with the position I take in this chapter: that transhumanism owes at least an originary debt to Nietzsche's thought. However, many of the regular contributors to *JET* took exception to this claim. Following Sorgner's essay, a 2010 issue of the journal was dedicated solely to debating the role of Nietzsche's thought in transhuman philosophy and rhetoric, presenting several short position papers, longer articles, and culminating in Sorgner's own response to his respondents.

As these responses make clear, there are a variety of transhumanisms, even within what could be recognized as the contemporary orthodox community. Despite this range of thought, a general consensus seems to be to reject Sorgner's assertion that Nietzsche is a transhumanist. Reasons for this insistence vary, but a common thread emerges most glaringly in William Sims Bainbridge's "Burglarizing Nietzsche's Tomb," a rather rambling account of Nietzsche's Apollonian/Dionysian distinction, Romantic Europe, and Nazism. Its introductory paragraph concludes, "Perhaps Nietzsche himself was the first transhumanist. Perhaps he really was a Nazi."[5], As Sorgner points out in his response, Bainbridge seems to be arguing that, despite the previous assertion, Nietzsche was a Nazi and that this Nazism means that Nietzsche *cannot* be a transhumanist (though, of course, as Sorgner points out, this is both ideologically problematic and historically impossible). Many of the readings of Nietzsche by Bainbridge's transhumanist peers, while not as explicitly depicting this concern for Nietzsche's contributions to Nazism, are wary of the philosophy of the

Übermensch due to the tainting horrors of National Socialism. As Sorgner carefully indicates in his response, this view of Nietzsche suggests that many transhumanists have not been keeping up with current scholarship on Nietzsche beyond the general revival in his thought started by Walter Kaufmann with the now-distant publication of *Nietzsche: Philosopher, Psychologist, Antichrist* in 1950.

Overall, the conversation between Sorgner and his respondents in this issue of *JET* is marked by missed connections, with Sorgner, in his response, continuing to reject the other contributors' assertions about Nietzsche. Sorgner takes recourse to "the current state of the art in Nietzsche scholarship" (especially obvious in his continued rejection of assertions that Nietzsche was antiscience).[6] It becomes clear throughout the debate that, despite not being familiar with the proliferation of studies dimensionalizing Nietzsche's thought over the past few decades, many of *JET*'s regular contributors remain emphatic that Nietzsche is *not* a transhumanist and that his thought has little to offer transhumanism.

Philosopher Max More's response to Sorgner, though, is an exception. Based on Sorgner's comments in his response, More's argument appears to be the only one with which Sorgner does not take issue—at least at the level of Nietzsche interpretation. More's most salient contribution to the debate in *JET*, as well as to my argument in this chapter, is that

> What we can infer is that differing variants of transhumanism are possible. Certainly there is no inconsistency between transhumanism and a utilitarian morality. But neither is there any inconsistency between transhumanism and a more Nietzschean view of morality. While Nietzsche viewed morality as essentially perspectival, we can easily enough fit him loosely within the virtue ethics approach classically represented by Aristotle. Yes, transhumanism can be sanitized and made safe so that it fits comfortably with utilitarian thinking. Or we can take seriously Nietzsche's determination to undertake a "revaluation of all values."[7]

More is right in pointing out that there are certainly many ways of getting at the essential evolutionary paradigm represented by transhumanism; documenting these paths is one of the main projects of this book, after all. However, I would like to make one key departure from More's account of the history of transhuman thought: I do not think the foundational idea of transhumanism—positing a limit to the human that must be overcome by the creation of some kind of posthuman (whether such a posthuman be informatic, cybernetic, or some other channel for overcoming)—is possible to imagine through utilitarianism. As with Bostrom claiming Gilgamesh as a fellow transhumanist, desires for improving the human condition have surely existed since the dawn of any kind of human record; however, transhumanism, if it is to refer to anything in particular, must be focused on the portion of evolutionary futurist argumentation at which the radical or alien comes out the other end of technologically mediated humanity. While the concept of using technology to improve human life is as old as history itself, the transhumanist idea that we will soon become unrecognizable to ourselves through our technology is not.

While such an evolutionary future is not implied in utilitarian philosophy, it is at the core of Nietzsche's understanding of the *Übermensch*: the being that will emerge after the limits of the human have been overcome. The frustrating aspect of this discussion, from the perspective of evolutionary futurism, is that many of the *JET* writers think they are discussing philosophical issues whereas Sorgner's original essay in 2009 attempts to draw a line from Nietzsche's thought to real-world scientific practice:

> Nietzsche upheld that the concept of the overhuman is the meaning of the earth. I think that the relevance of the posthuman can only be fully appreciated if one acknowledges that its ultimate foundation is that it gives meaning to scientifically minded people.[8]

Sorgner here suggests that Nietzsche's concept of the *Übermensch* provides an imaginative foundation for the entire project of actualizing a posthumanity—which is, of course, one of the generally agreed upon goals unifying transhumanists across the world. Contra More, the question is not of what moral position a transhumanist might take (Nietzschean or utilitarian), but of the ideas that made other later knowledge formations possible. We can imagine a Nietzschean or a utilitarian ethical stance for a transhumanist to take, but the idea of a future-as-radical-break is not possible without Nietzsche. In quibbling over philosophical issues between this or that system, the point that Nietzsche inaugurated the idea of the human as a limit to be overcome gets lost.

To restore and intensify this concept's originary insight for transhumanism, I now turn to explore two aspects of transhuman rhetoric through the anxiety of Nietzsche's influence on the discourse. I examine the origin of evolutionary futurist rhetoric in the culture of avant-garde modernist movements born in the confluence of Nietzsche and Henri Bergson's creative evolution in early twentieth-century Europe. Drawing from Mina Loy's writings on feminism and futurism and P. D. Ouspensky's mystical speculations on Darwin and the *Übermensch,* I trace an early thread of transhuman rhetoric chiefly concerned with *topoi* of cosmic consciousness and cognitive evolution. From these modernist origins, I trace how this trope of mental evolution drops in and out of evolutionary futurist practice throughout the twentieth century. This flickering ultimately establishes a dialectical relationship between the mind and the body in discussions of transhumanism, one that is increasingly important given the political and economic stakes associated with the commodification of cognition as outlined in the introduction.

Modernism's Transhumanism?

As I suggested above, my understanding of tranhumanism's modernist origins hinges primarily on Nietzsche's concept of the *Übermensch.* In constructing the human as a limit to be overcome, regardless of how specific overcomings may manifest themselves in various transhuman philosophical programs, Nietzsche's core insight cannot be overestimated.[9] In further making this case, I explore two specific modernist artist/philosophers who both, on encountering Nietzsche in their wanderings, almost immediately began producing work that conforms to the patterns of evolutionary futurist rhetoric. I first discuss Mina Loy's contributions to the Futurist movement and her deep commitment to evolutionary futurism before switching to P. D. Ouspensky's singular commentary on Nietzsche, in which Ouspensky uses the figure of the *Übermensch* to square discourses of magical philosophy with Darwinian evolution. In both cases, we see how contact with the idea of the human-as-limit produces an immediate and specifically evolutionary futurist graphomania.

Cosmic Consciousness, Futurism, "Woman": Mina Loy's Transhumanism

Janet Lyon's excellent account of Mina Loy's avant-garde feminism introduces the poet, playwright, manifesto writer, futurist, and painter with the following biographical gloss, tracing Loy's response to the trope of irrational female chaos in the writings of her male futurist counterparts (especially F. T. Marinetti):

> It stretches through several of the works that she produced while in Florence, where she lied with and was then estranged from her English husband, where she bore three children and lost one in infancy, where she cultivated

a friendship with the Americans Mabel Dodge, Gertrude Stein, and Carl van Vechten, and where she exhibited paintings at the First Free Exhibition of International Futurist Art. These were the years when she experienced what she called "the throes of conversion to Futurism." She allied herself with the iconoclastic energy of futurist aesthetics and—just as important for her critique of futurism—had affairs with Marinetti and Giovanni Papini, the political editor of *Lacerba*.[10]

Initially swayed by the eugenic, hygienic approach to futurity projected by the early days of the futurist movement in Italy, Loy became critical of the model of republican motherhood being advocated in tracts such as Valentine de Saint-Point's "Manifesto of Futurist Women," in which the role of women was to birth and foster male genius.[11] As Lyon continues, providing further biographical background,

> By the end of 1914, after she had been through her sexually and emotionally disappointing affairs with Marinetti and Papini, her enthusiastic echoes of futurism gave way to probing analyses of futurism's platform concerning women and heterosexuality. From 1914 to 1916 she produced a number of remarkable critiques of futurism's relation to Woman and futurists' relations with women, including the unpublished "Feminist Manifesto" (1914), the poem sequence *Songs to Joannes* (published 1915), the unpublished play *The Sacred Prostitute* (ca. November 1914), and the play *The Pamperers* (1916; published 1920).[12]

This canon of futurist texts (to which I will add "Parturition" [1914]) observes with a sharp critical eye the construction of femininity within the futurist avant-garde during the years leading up to World War I.

As Lyon explains, Loy's critique of futurism probes

> the Manichaean rift between the sexes wrought by futurists and their continental progenitors. In them, as in a host of other works, Loy plays with futurism's taxonomical constructions of "woman" and extemporizes her own alternative taxonomies of "man": she also highlights deferred spaces of meaning between "woman" and "man" and so unhinges futurist certitude about the ontologically gendered foundations of avant-garde poetics. Finally, throughout her work Loy simply refuses to give up any of her own claims to avant-garde authenticity.[13]

In Lyon's reading of Loy, this criticism of a simplistic and profoundly discriminatory gender system (in which women are constructed in fragmented opposition to a philosophically smooth and whole masculinity endowed with action and speed) hinges on articulating a similarly fragmented model of masculinity. Further, Lyon intimates that Loy's criticism of futurism stands in opposition to the specifically Nietzschean fervor of futurism's overcoming.[14]

It is this claim I want to complicate in this section. In my reading of Loy's rhetoric, I find that her writings of critical futurism strongly deploy evolutionary futurist tropes and patterns. Moreover, Loy's writing is strongly supportive of the Nietzschean posthuman paradigm. Lyon argues that Loy is critical of Nietzsche because of the predominant misunderstanding of what is meant by *Übermensch.* However, as Carolyn Burke makes clear in her biography, *Becoming Modern: The Life of Mina Loy* (1997), Loy read widely in Nietzsche and others in the "New Thought" whose wider circulation in Europe marked a beginning of avant-garde modernism.[15] As additional evidence of her affinity for Nietzschean thought, in "Aphorisms on Futurism" (1914), Loy borrows the form of Nietzsche's late writing (the aphorism) to make her most orthodox contribution to futurism's manifesto-driven culture. In fact, Burke's biography mentions a specific episode in 1903 in which, tellingly, Loy dismisses the occult dabbling of Aleister Crowley and other occultists for having only, as Burke puts it, "slight knowledge of Bergson, tinged with an even slighter appreciation of Nietzsche."[16] As Loy wrote in the manuscript for her never-published autobiography, from which Burke cites throughout *Becoming Modern,* she was unsettled by "their somewhat sinister conviction of being supermannish"—less a rejection of Nietzsche per se than a disappointment with bad Nietzscheanism that she would level against the futurists later in her life, as I discuss below.

This misunderstanding of Loy's engagement with Nietzsche may hinge on finding a translation for the German phrase that accurately captures Nietzsche's multivalent meaning for *Übermensch.* Many early translators chose to render the phrase as "superman" in English. As Michael Tanner declares in *Nietzsche: A Very Short Introduction,* "I find 'superman' absurd," mirroring the opinion of many philosophers who grew up in a culture familiar with *Action Comics.*[17] Beyond mere embarrassment, though, nearly every book about Nietzsche feels some need to comment on the chosen translation of the German phrase. Ullrich Haase's account, from *Starting with Nietzsche,* offers a useful account of the stakes involved in the particular translation of this concept:

> The notion of *Übermensch* in Nietzsche's works has long attracted considerable interest and it is perhaps not too surprising that much of this interest has led to sometimes amusing and sometimes catastrophic interpretations. Thus in particular the National Socialists of the German Third Reich have made out of this *Übermensch* the caricature of a self-willed "Blond Beast." The first translation of this term in Nietzsche's texts as the *"Superman"* has only worsened this absurd image. This idea of the *Superman* "having" a great *Will to Power* by means of which he would subjugate other humans has, unfortunately, held sway for a long time. This is not on account of any close reading of his works, but simply because that is how one could easily understand the terms *Will to Power* and *Superman*. Consequently, many translators have adopted the more literal translation "Overman." This translation makes more sense in terms of the *Über*, but still suffers from implying a single individual. But Nietzsche does not speak about an individual male or female human being, but about the historical existence of the "human being."[18]

Haase concludes by stating that he translates *Übermensch* as "Overhuman" because it captures the essence of Nietzsche's original concept: the *Übermensch* replaces the human as one philosophical concept replaces another, not as one individual replaces another. Far from being an actual, existing figure (as it is often documented in science fiction and comic book appropriations), the *Übermensch* is a guidepost for becoming: a beacon to humans who want to overcome the contingent and base existence Nietzsche associates with humanist conceptions of life.

So much of our rhetorical associations with the concept of *Übermensch* invoke individual superbeings (who often wear red capes), but how else can the *Übermensch* be a guidepost? The *Übermensch* represents an idea of a future being that is beyond the human condition. This idea does not have specific characteristics. Instead, the *Übermensch* stands for all avenues beyond the human, without a specific shape. Ullrich Haase uses a specifically ingenious means of explaining the relationship between *Mensch* and *Übermensch*: he imagines an ape thinking about a future in which their species ends. While the ape will not be able to articulate the specific shape of the species to come, it may be able to name this species to come "the overape." Of course, in retrospect, we can see that the overape is the human, but we cannot know that before the fact.[19] In other words, "*Übermensch*" is what we call any being that will come to fill our place as the dominant species on Earth.

This distinction between individual supermen and a coming overhumanity is critical to understanding both Loy's critique of futurism *and* her importance in building a conceptual bridge between avant-garde modernism and transhumanism. Loy is not critical of Nietzsche qua Nietzsche. Instead, she is critical of what we might call the "sullen little boy" model of Nietzsche, in which (and I am assuming we have all encountered these kinds of people) a teenager reads Nietzsche and assumes, because of his[20] perceived intellect, that he *is* the *Übermensch*. This is the sort of bad reading of Nietzsche that Loy assigns to the futurists. As Lyon writes, for Loy "no amount of individual will . . . can wish away the political, social, and biological components that accrue to gendered subjectivity."[21] The often absurdly hyperbolic position of futurists such as Marinetti becomes petty teenage bombast in the face of the messy gendered realities Loy documents, plastering over a frightening, confusing reality with a machismo masking the scared and lonely teenager within.

Loy's distinction is drawn out in Lyon's reading of the dichotomy between lust in futurist writing and sex in Loy's transhuman poetics. In Lyon's reading, lust (which is the watchword in Valentine de Saint-Point's various manifestos on futurism and women) is merely a "unilateral drive," the kind of naive will-to-power of the sullen and solitary teenage Nietzschean. In contrast, Lyon argues, "'sex' constitutes a rare overlap of subjectivities"; it is "intersubjectivity," "an even more dynamic composition of both bodies and psyches."[22] In this reading, Loy's reconstitution of gender merges with the merging of bodies and souls in the congress of sex to open new vistas of personal evolution as well as species evolution. Lyon concludes her observations with this intersubjective understanding of eros, but I want to push on these evolutionary vistas. Regarding Loy's accounts of sexuality and her readings of the *Übermensch* as signposts to be worked toward can, I argue, contextualize her writing as the first major instantiation of a particular evolutionary futurist *topos*, which psychologist Richard Bucke labeled in 1901 as "cosmic consciousness."

Bucke, a former asylum director, captures in *Cosmic Consciousness: A Study in the Evolution of the Human Mind* (1901) an evolutionary model of consciousness in which various stages are dynamically and evolutionarily explored by life in its various configurations. His formulation suggests that there exist two widespread forms of consciousness: the simple consciousness of animals and the reflective consciousness of humans (which he calls "self consciousness").[23] In this formulation, human consciousness is singular, and singularly evolutionary, because of its ability to reflect back on itself. For Bucke, if we pause and meditate, we can move our consciousness to higher orders of contemplation. He finds models for this evolutionary thought in the life of Buddha and in Jesus. This reflexive quality is, beyond Bucke, central to many of the transhuman accounts of the evolution of consciousness. For instance, Pierre Teilhard de Chardin praises "the spiritual phenomenon of reflection" in his outline of noöspheric evolution, which I will discuss in greater depth in chapter 3. Similarly, sourcing his philosophy from Albert North Whitehead, Alfred Korzybski, who devised a system for overcoming the human through manipulations of basic grammatical rules, posits reflexivity as the chief means by which thought may become thought about thought, thereby opening the path to higher consciousness. Additionally, we find this idea of reflexive cognitive evolution at the core of a variety of mystical and cybernetic systems, including Timothy Leary's multifaceted theory of cybernetic circuits of mind outlined in *Info-Psychology,* the Spiral Dynamics of Beck and Cowan, the concept of *aša* in Zoroastrianism (itself an inspiration for Pythagoras's concept of Cosmos in the first place), and the Integral Yoga of Sri Aurobindo. This tropic treatment of human consciousness, popularized in the West in the twentieth century by Bucke, has long been central to an evolutionary futurist rhetoric.

More directly relevant to Mina Loy's poetics and more connected to evolutionary futurism, however, is Bucke's third stage of consciousness evolution, what he labels "cosmic consciousness." Cosmic consciousness is an enlightened modality of thought, which Bucke says is "a consciousness of the cosmos, that is, of the life and order of the universe."[24] Bucke states that this third state of consciousness is accompanied by "an intellectual enlightenment or illumination which alone would place the individual on a new plane of existence," along with "an indescribable feeling of elevation, elation, and joyousness, and a quickening of the moral sense."[25] Cosmic consciousness becomes a state of being at peace with the entirety of existence, in which a "universal scheme is woven in one piece and is permeable to consciousness."[26] This cosmic frame of mind—beyond intersubjectivity—is the model foregrounded in Loy's writing as she moves beyond futurism. For Lyon, Loy's poetics atomize masculinity in the same way that sexist discourse in futurism atomizes (and thereby subordinates) "the" "feminine." Loy is *also* doing something else, something cosmic. In her "Aphorisms on Futurism," a text Lyon describes as being primarily involved in "the eugenic individualism of the futurist mode," Loy begins to make the case for cosmic consciousness as being of interest to (and perhaps of paramount importance for) the corporeally obsessed (male) futurists.[27] Loy writes,

> TODAY is the crisis in consciousness.
> CONSCIOUSNESS cannot spontaneously accept or reject new forms, as offered by creative genius; it is the new form, for however great a period of time it may remain a mere irritant—that molds consciousness to the necessary amplitude for holding it.
> CONSCIOUSNESS has no climax.
> LET the Universe flow into your consciousness, there is no limit to its capacity, nothing that it shall not re-create.
> UNSCREW your capability of absorption and grasp the elements of Life—*Whole.*[28]

Contrast this with F. T. Marinetti's statement on the absolute in "The Founding and Manifesto of Futurism" (1909):

> We stand on the last promontory of the centuries! . . . Why should we look back, when what we want is to break down the mysterious doors of the Impossible? Time and Space died yesterday. We already live in the absolute, because we have created eternal, omnipresent speed.[29]

Marinetti's manifesto, inaugurating futurism, is obsessed with death, destruction, and liberating Italy from "the smelly gangrene of professors, archaeologists, *ciceroni* and antiquarians."[30] Through the speed of the motorcar (which propels Marinetti and his futurist colleagues on the mad chase after Death that famously opens the manifesto), humanity has already unlocked the absolute, the limits of human being. The only task remaining for art is to destroy the past in order to liberate this aesthetic of speed so that humanity can wake up to a new machinic consciousness.

In contrast to this violent car crash with the absolute of the techno-modern death drive, Loy's position in "Aphorisms on Futurism" is radically different: radically more transhuman, and, generally, more radical. Unlike Marinetti's boys in fast cars, Loy recognizes (as Bucke similarly argues) that humanity exists on a continuum of cognitive evolution that, importantly, "has no climax." For Loy, the human is a transitory configuration in a much longer unfolding of consciousness into cosmos, an "unscrew"-ing of humanity's "capability of absorption" to "grasp" the "*Whole.*" This merging and expanding imagery is quintessentially transhuman and has little to do with the violent individualism of Loy's literally small-minded peers in the futurist movement.

Loy's transhumanization and the cosmic consciousness that results continues to develop as she more fully frees herself from the trappings of Italian futurism. In many ways, from our later historical perspective, Mina Loy is a transhumanist in search of a discourse. While futurism initially seems to provide this for her (especially in the "Manifesto" and "Aphorisms"), as she increasingly regards futurism as the boys-with-toys club that it (mostly) was, she begins to actualize her own evolutionary futurism, marking her as an early transhumanist. Beyond the allusions to a cosmic consciousness in "Aphorisms on Futurism," the theme begins to explicitly appear in Loy's work in "Parturition," an account of giving birth in which Loy's experience as "the centre / Of a circle of pain" opens up vistas to the kind of totality Bucke described.[31] Moreover, Loy describes this experience as "cosmic," first referring to her birthing chamber as "my congested cosmos of agony" before later describing

> A moment
> Being realization
> Can
> Vitalized by cosmic initiation
> Furnish an adequate apology
> For the objective
> Agglomeration of activities
> Of a life
> LIFE
> A leap with nature
> Into the essence
> Of unpredicted Maternity[32]

In this section, she describes "consciousness in crises" racing "Through the subliminal deposits of evolutionary / processes" on the way to an experience of what, in certain circles, would be described as ego death. Transpersonal psychologist Stanislav Grof clarifies ego death as,

> When experienced in its final and most complete form, the ego death means an irreversible end to one's philosophical identification with what Alan Watts called *skin-encapsulated ego*. When the experience is well integrated, it results not only in increased ability to enjoy existence, but also in better functioning in the world. The experience of total annihilation and of "hitting the cosmic bottom" that characterizes the ego death is immediately followed by visions of blinding white or golden light of supernatural radiance and beauty.

> It can be associated with astonishing displays of divine archetypal entities, rainbow spectra, intricate peacock designs, or pristine natural scenery. The subject experiences a deep sense of spiritual liberation, redemption, and salvation.[33]

Grof's description fairly accurately captures the narrative trajectory of Loy's poem, moving from the pain of her "congested cosmos of agony" to a fuller experience of being:

> Stir of incipient life
> Precipitating into me
> The contents of the universe
> Mother I am
> Identical
> With infinite Maternity
>   Indivisible
>   Acutely
>   I am absorbed
>   Into
> The was—is—ever—shall—be
> Of cosmic reproductivity[34]

In this moment, Loy finds her suffering transmuting into an experience of the cosmos as totality, the realization that her act of birthing connects her to a universal, transhistorical cycle of birth and death. She follows this moment with two episodes rising "from the subconscious."[35] The first is a cat giving birth outside her apartment, surrounded by the "Same undulating life-stir"; Loy concludes "I am that cat."[36] The second experience is of "Impression of small animal carcass / Covered with blue bottles" that, because of the feasting flies, "Waves that same undulation of living."[37] In both cases, living and dying, Loy realizes that her experience of birth connects her to a larger force than herself, what Henri Bergson might label "élan vital," a cosmically unified being (in addition to providing the "pristine natural scenery" Grof notes as being associated with experiences of ego death). This experience Loy narrates is the break from "*skin-encapsulated ego*" as the burst into "spiritual liberation" beyond egoic individuality.

I iterate on this idea of cosmic consciousness in Loy's poetry because Bucke refers to the process of attaining cosmic consciousness as a process by which we must "transhumanize." As Bucke suggests ("this is what is called in the East the 'Brahmic Splendor,' which is in Dante's phrase capable of transhumanizing a man into a god"), the most ancient documented use of "transhuman" occurs in Canto I of Dante's *Paradiso,* where it is "*transumanar*" in the original Italian. R. Allen Shoaf glosses this curious neologism as referring to the need for Dante's narrator to undergo a series of morphological changes in his vision to experience the divinity of Heaven directly and then to convey it to his readers.[38] It is in this divine morphological manner that T. S. Eliot also uses the verb in *The Cocktail Party* (1949), where it refers to the suffering of a soul on its way to a more complete experience of divinity.[39] Although this term obviously shares etymological roots with "transhumanism," it is used in a slightly different way in Bucke and Dante than it is in Max More or Bostrom. However, in exploring more fully this earlier use of the term "transhuman" (as a verb), the link between Loy, Bucke, and transhumanism becomes more apparent. Such a reading is, importantly, not to argue that Bucke or Dante are transhumanists (though there are a number of forum and blog posts attempting to argue this very point based on Dante's usage). Instead, Dante's usage as a kind of mutation of the body for the better contemplation of the divine provides a thread of shared thought between the mystical overcoming of human cognitive limits and the transhuman overcoming of human physical ones. In its explicitly religious usage, the verb "transhumanize" refers to advancing human morphology so that our limited sensorium may contemplate the divine. It would not be until the work of FM-2030 in the 1970s that the noun form ("transhuman") or the ideology ("transhumanism") emerged fully into our lexicon. Although contemporary transhumanism—with its emphasis on rationalism, scientific proof, and bureaucratic planning—places a strong emphasis on reason, the link to Dante through Bucke and Loy helps to explain why transhumanism is so often discussed in millenarian terms and why the movement so often ends up articulating a kind of cybernetic mysticism.

In any case, Bucke's work is significant for synthesizing Western and Eastern concepts of attaining divinity and articulating a potent set of concepts and *topoi* for what would become New Age thought. However, for an emergent evolutionary futurism, his specific revision and placement of these concepts within a secular framework of psychology is perhaps even more important. Additionally, his use of *cosmic* to capture this mystical-scientific concept is telling. As I argued in the Introduction, a major contributing factor to evolutionary futurism is the insistence that technological change provokes biological evolution. By thinking with

Bucke, we can see that the kinds of technology often signaled as exerting these pressures are, in his model, cosmic in nature. The next step up from consciousness of consciousness is, for Bucke and like-minded thinkers, consciousness of the totality of all: the cosmos. According to Alexander von Humboldt, author of *Kosmos* (1843)—the work that first returned the more expansive meaning of cosmos to our lexicon—in Ancient Greek,

> *κόσμος*, in the most ancient, and at the same time most precise, definition of the word, signified *ornament* (as an adornment for a man, a woman, or a horse); taken figuratively for *εὐταξία* [meaning order, discipline, or method], it implied the order or adornment of a discourse. According to the testimony of all the ancients, it was Pythagoras who first used the word to designate the order in the universe, and the universe itself.[40]

"*Cosmos*," at the time of von Humboldt, was often thought of and translated as a kind of rhetorical style or adornment. However, by tracing the Pythagorean legacy of this term, he recovers the meaning for the term used by Bucke: the order of all that is. By linking Dante's "transhumanize" with this understanding of cosmos, Bucke inaugurates an important rhetorical figure in evolutionary futurism: the idea that an evolution in consciousness yields an expanded awareness of the all. For Loy, the pain of childbirth inaugurates this experience in her writing, but in treating the figure of the machine, so important to the Italian futurists, she begins to imagine truly startling, recognizably transhuman visions of our evolutionary future. The tropic intensification of Mina Loy's transhuman poetics continues in her long cycle of thirty-four poems, *Songs to Joannes* (1917). As Lyon observes, the poem uses the cosmic perspective Loy experimented with in "Parturition": "in a series of short irregular poems, we are shown the fragments of a discontinuous sexual and intellectual relationship; but we are also shown alternative fragments, equally discontinuous, of a relationship that *might* have been."[41] These interlocking perspectival fragments force us to confront the true nature of cosmic consciousness: "that 'what is' is only one of several sets of charged fragments, the presence of which undermines any definitive claims to comprehensive representation."[42] *Songs* creates a shifting crystalline network of poems documenting a distinctly Pythagorean version of cosmos: literally all that can or could or did or will or might exist. As Lyon makes clear, this cosmic perspective is opened up through Loy's usage of sex as a fragmentary, transpersonal becoming to counteract the masculine, linear, penetrative *topoi* of lust in orthodox futurist thought.[43] Rather than the temporary experience of cosmos described in "Parturition," *Songs* is a continued evocation of a higher form of consciousness.

Animated by futurism's tropic matrix, though highly critical of its methodologies, Loy's collection of poems begins to manifest imagery of this cosmic evolution in terms that, in a contemporary context, can be read as recognizably transhuman. In Song 25, she imagines a group of characters becoming "machines . . . cutting our foot-hold / With steel eyes."[44] Later, in Song 29, Loy imagines the possible outcome of these newly machinic human beings:

> Evolution fall foul of
> Sexual equality
> Prettily miscalculate
> Similitude
>
> Unnatural selection
> Breed such sons and daughters
> As shall jibber at each other
> Uninterpretable cryptonyms
> Under the moon[45]

These new products are the result of "unnatural selection," of something other than an evolution that implies sexual inequality. These "sons and daughters" grow up to "jibber" in "[u]ninterpretable cryptonyms," which speaks to a specifically Nietzschean understanding of cosmic consciousness. Just as Nietzsche describes the *Übermensch* as "the lightning from the dark cloud of 'human being,'" these offspring of an unnatural evolution remain cryptic to us: unknowable and untranslatable.[46] The remainder of Song 29 educates these products of "unnatural selection" to develop differently and not fall foul of human emotional suffering. This poem concludes while these children "clash together . . . / In seismic orgasm" and ends with the final failure of recognition on the part of parents: a "[w]ince in the alien ego."[47] Thus Loy shifts the tropic landscape of futurism from the red-cape-wearing individual *Übermensch* to a model of a coming overhumanity—one similar to contemporary interpretations of Nietzsche, suggesting Loy to be one of the foremost readers of Nietzsche during this period. In her poetic experimentation with cosmic transhumanization *and* the machinic trappings of a rejected hypermasculine futurist art, Loy begins to craft a vision of becoming that is mediated through technology and focused on the refinement of thought itself, the kind of inner transhumanism I am tracing in the modernist avant-garde during this period. In the next section, I show how P. D. Ouspensky intensifies this claim while inserting an inner transhumanism into the discourse of Darwinian evolution, documenting the cognitive stakes involved, from a biological perspective, with actualizing any kind of *Übermensch*.

Mysticism, Psychology, Darwinism: P. D. Ouspensky's Transhumanism

While Mina Loy's avant-garde poetics traces an evolutionary futurist trajectory for Italian futurism through a synthesis of Nietzsche's concept of the *Übermensch* and Bucke's concept of cosmic consciousness, the commentary on Nietzsche offered in P. D. Ouspensky's *New Model of the Universe* (1917) further intensifies the connection between the mystical experiences suggested in Bucke and the idea of an evolutionary futurism for humanity. A mathematician by training, P. D. Ouspensky was also a practicing magus and disciple of G. I. Gurdjieff, a teacher who articulated a radical version of Perennialism. Ouspensky, one of Gurdjieff's most devoted disciples, was responsible for introducing Gurdjieff's work to England in the 1920s, as well as directly introducing the system to a number of modernist authors and artists.[48] Moreover, his commentary on Nietzsche was important to an emerging evolutionary futurist sensibility, namely, through Ouspensky's synthesis of this magical milieu and Darwinian discourses of evolution. Ouspensky, even more than Loy, reveals the mystical impulses at the core of transhuman thought and suggests profound ways of rethinking the fringe status of the contemporary movement, especially from a rhetorical context.

This commentary on Nietzsche is part of the larger argument traced by *New Model of the Universe*. In his preface to the second edition, Ouspensky introduces the collection as being about esotericism and evolution:

> The idea of esotericism is chiefly the idea of *higher mind*. To see clearly what this means we must first of all realize that our ordinary mind (including the mind of a genius) is not the highest possible order of human mind. The human mind can rise to a level almost inconceivable for us, and we can see the results of the work of higher mind . . . in the Upanishads, in the Mahabharata; in works of art such as the Great Sphinx of Gizeh, and in other memorials though they are few in literature and art.[49]

As he unpacks the whole collection, this higher mind (analogous to "cosmic consciousness" in Bucke) results from training in esoteric and mystical methods. Additionally, this higher mind is, as he explores through Nietzschean themes, specifically evolutionary in nature. However, where Bucke uses evolution as a method in which higher order is extrapolated from lower, Ouspensky means to argue that the move toward the *Übermensch* is evolutionary in the biological specifics discussed by Charles Darwin.

Ouspensky's commentary on the *Übermensch* explores two major avenues: genealogy and synthesis. In the first of these two threads, he historicizes Nietzsche's concept within a longer lineage of Jungian archetypal heroes and villains. The second thread considers the role of magic as a proto-transhumanist evolutionary method in the face of the reductive and increasingly dominant paradigm of Cartesian science. Both of these threads are important to understanding the status of evolutionary futurism in modernism, as the *topoi* of the modernist avant-garde begin to mutate into recognizably transhuman formations in the work of Ouspensky. In this section, I will briefly unpack Ouspensky's genealogy of the *Übermensch* before exploring, in more depth, his construction of magic as a rhetorical, Utopian method for actualizing transhuman overcoming.

For Ouspensky, Nietzsche's concept of the *Übermensch* is a tool for overcoming the flattening and simplifying Cartesian humanist framework—in other words, the framework that has successfully asserted that the proper object of knowledge production and philosophizing is "man as he is, as he always was and always will be."[50] In contrast, Ouspensky claims that the *Übermensch* represents a rhetorical break that is "never satisfied with man as he is," an understanding that is both a rhetoric of "the masses" and a holdover from pre-Enlightenment philosophies.[51] As we will see below, this premodern trace enables Ouspensky to claim magic as a preeminently transhuman methodology.

To this end, Ouspensky's genealogy of the *Übermensch* concept in Nietzsche begins by asserting that "the idea of superman is as old as the world. Through all the centuries, through hundreds of centuries of its history, humanity has lived with the idea of superman."[52] Before the onset of a logical, rational human framework, the *Übermensch* was a mythical figure associated with "a legendary Golden Age," in which "life was governed by supermen, who struggled against evil, upheld justice and acted as mediators between men and the Deity, governing them according to the will of the Deity, giving them laws, bringing them commandments."[53] Alluding to figures such as Gilgamesh, King Arthur, and Odysseus, Ouspensky suggests two mutations to this mythic *Übermensch* following the start of the Enlightenment. On the one hand, the mythic hero persists in popular adventure literature ("what indeed is the Count of Monte Cristo, or Rocambole, or Sherlock Holmes, but a modern expression of the same idea of a strong, powerful being, against whom ordinary men cannot fight . . . ?") but also in the kind of neutered futurism Nietzsche associated with Christian salvation history.[54] In this latter mode, the *Übermensch* "was to come, arrange their affairs, govern them, teach them to obey the law, or bring them a new law, a new teaching, a new knowledge, a new truth, a new revelation. The superman was to come to save men from themselves, as well as from the evil forces surrounding them."[55] As lawgiver and order maker, "the image of superman in this case loses all colour and grows almost repulsive, as though from the very fact of becoming lawful and inevitable."[56] For Ouspensky, this model of *Übermensch* as savior represents the corrosive quality of Cartesian humanism.

In contrast to this passive model of futurity, in which humanity must only await rescue by a future *Übermensch*, Ouspensky praises the active, adventurous *Übermensch* of popular fiction. This component of adventure proves the linchpin to Ouspensky's second point, which concerns magic—not science—as the method for actualizing an overcoming of the human. Specifically, Ouspensky's account of magic is tied to an understanding of evolution that significantly and importantly complicates simple understandings of an evolutionary *telos*. Specifically, he is critical of how "evolutionary theories have become the basis of a naive optimistic view of life and of man," coupled to the colorless *Übermensch* bearing a passive future history of salvation.[57] He continues: "It is as though people said to themselves: now that evolution exists and now that science recognizes evolution, it follows that all is well and must in future become still better."[58] Ouspensky recognizes the importance of Darwinian evolution for the *Übermensch* but understands that this mechanism still presents a problem. As R. J. Hollingdale puts it in his biography of Nietzsche,

> natural selection was for Nietzsche essentially evolution freed from every metaphysical implication: before Darwin's simple but fundamental discovery it had been difficult to deny that the world seemed to be following some course laid down by a directing agency; after it, the necessity for such a directing agency disappeared, and what seemed to be order could be explained as random change.[59]

For Ouspensky, the reception of this "random change" has led to another evolutionary passivity: a scientific reinscription of religious salvation history in which natural selection replaces God as the mechanism by which the savior *Übermensch* arrives in the future. Under this understanding of biological evolution as destiny, humanity must, again, merely await its inevitable salvation.

Ouspensky, reminding us of this danger, instead suggests that Darwinian evolution *must* be understood as risk:

> Evolution, however it be understood, is not assured for anyone or for anything. The theory of evolution means only that nothing stands still, nothing remains as it was, everything inevitably goes either up or down, but not at all necessarily up; to think that everything necessarily goes up—this is the most fantastic conception of the possibilities of evolution.[60]

Given this up-and-down character, evolutionary change is not inevitable. Instead, Ouspensky spends much of his commentary making the case for removing both evolutionary positivism and the passive logic of salvation history from discussions of the *Übermensch*. Evolution, for Ouspensky, is something humans make for themselves only through blind experimentation.

This emphasis on experimentation reveals both Ouspensky's nascent transhumanism and his divergence from contemporary transhuman thought. On the one hand, both systems share an understanding that because of intelligence, humanity is now driving its own evolution. However, for Ouspensky there is no one path toward this next stage in evolution; the path to the *Übermensch* will necessarily be littered with failed experiments. As he explains,

> The evolution of consciousness, the inner growth of man, is the "ascent towards superman." But inner growth proceeds not along one line, but along several lines simultaneously. These lines must be established and determined, because mingled with them are many deceptive, false ways, which lead man aside, turn him backward or bring him into blind alleys.[61]

For Ouspensky, the process of becoming transhuman, of evolutionary futurism itself, lies in mapping out these various pathways. These experiments, however, are not into the scientific qualities of mystical phenomena. Instead, "the development of the inner world, the evolution of consciousness, this is an absolute value, which in the world known to us can develop only in man and cannot develop apart from him."[62] Here Ouspensky is arguing for an experimental magic that actualizes human evolutionary change.

In building this case, Ouspensky makes an extended argument that a theory of evolution rooted in the traditions of magic, rather than in the traditions of science, will be the best methodology for this experimentation. In aligning his methodology against science, Ouspensky participates in a larger discussion percolating into various facets of European culture at the turn of the century. Ouspensky encapsulates this conversation:

> The literature on magic and occultism was for a long time entirely ignored by Western scientific and philosophical thought or rejected as an absurdity and a superstition. And it is only quite recently that people are beginning to understand that all these teachings must be taken in a symbolical way, as a complex and subtle picture of psychological and cosmic relations.

However, this reanimation of magic may, in fact, be unclear even to those of us who live ostensibly on the other side of this return.

As rhetorician William Covino observes in making the case for a return of "magical rhetoric" at the core of postmodern rhetorical theory, for most of us, "magic means the inexplicable and spontaneous materialization of a finished product; this is the familiar rabbit-out-of-a-hat definition."[63] Instead, Covino offers "an alternate definition, grounded in anthropological and sociological conceptions of magic."[64] This alternate definition reverses the common understanding of magic, suggesting instead that

> magic is not the instant and arhetorical product of an otherworldly incantation; it is the process of inducing belief and creating community with reference to the dynamics of a rhetorical situation. Magic is a social act whose medium is persuasive discourse, and so it must entail the complexities of social interaction, invention, communication, and composition.[65]

Covino's anthropological definition allies with Ouspensky, and thinkers such as Ouspensky and his teacher Gurdjieff spent their lives trying to return this definition of magic to a position of cultural prominence.

Additionally, Ouspensky's substitution of magic for science is embedded within a much more complicated conversation around the nature of magical or, more broadly, occult phenomena such as telepathy and ghosts during the dawn of modernism. Documented in Leigh Wilson's *Modernism and Magic* (2012), this debate was important for shaping modernist avant-garde practices for a number of reasons, especially given that modernist usage of mimesis is inherently magical in conception. This relationship between mimesis and magic emerges in James Frazer's hugely influential text *The Golden Bough* (1890–1915), in which magic "assumes that the same powers are present in the material objects once in contact with a person (a lock of hair, nail clippings) as they are in obvious copies of the person—the doll, for example. The repetition of the materiality of the world *is* mimesis, but a mimesis which . . . mistakes representation for the real thing, the ideal for the real."[66] This association is important for modernists, Wilson argues, because with the magical mimesis that fascinated the modernists, "if the copy is *more* than the original . . . that *more* is precisely and only the formal work which has produced it as a copy."[67]

Wilson suggests that for the staunchly Victorian Frazer, unlike the modernist avant-garde who used his work to other ends, this mimetic relationship is why magic fails. He writes that "the fatal flaw of magic is not in its general assumption of a sequence of events determined by law, but in its total misconception of the nature of the particular laws which govern the sequence."[68] Wilson diagnoses the causes of this failure by clarifying that, "while magic is like science in that it is a systematic form of thinking, where it differs is in its incapacity in testing its assumptions."[69] This distinction opens a discussion of the Society for Psychical Research, a society dedicated to the serious study of paranormal phenomena that included as a member Henry James and, in 1913, was presided over by Henri Bergson.[70] The Society, departing from Ouspensky's method, sought to extract experimental results to "reveal occult phenomena as part of the laws of nature as they were being revealed by science."[71]

As mentioned, this legitimation of occult research through experimentation is at odds with what Ouspensky is doing with Nietzsche's concept of the *Übermensch*. Ouspensky uses magic for the generative association that rhetoric and magic held prior to the Enlightenment. As George Kennedy summarized (and as anyone familiar with the history of rhetoric no doubt knows),"for some two centuries rhetoric made a claim to be the queen of the arts" in Renaissance Italy.[72] This regal importance derived, as Covino argues, out of the fact that, from the Classical era "through the Renaissance, words possess *actual* (rather than *symbolic*) power as agents of magic, and their effects are understood to vary with changing contexts."[73] This understanding of rhetorical force through literal word power is, as Covino argues, due to the magical tradition inherited from the Classical period. In contrast to understanding magic and rhetoric as entwined power acts that work by manipulating the social, the "mechanical universe issuing from the Enlightenment" imagines that "mind exists apart from matter."[74] This mind/body split, beyond being the foundational gesture of any Cartesianism, is inherently antimagical, and as Covino argues, antirhetorical.

For Covino, this antimagical, anti-Classical science sought to replace a world in which action was a rhetorical practice with the "clear observation language" needed for both empirical science and logical positivism.[75] Like the Society for Psychical Research that tried to subject magic to scientific practices of verification and truth, Covino argues that the revival of Classical rhetoric represented by the current-traditional turn in rhetorical studies[76] defangs this magical understanding of rhetor as magus who "must align the elements and the right words and the paths of the stars."[77] In both cases, what results are systems that do not embrace the full and radical potentials originally invoked by the discourses they tame. For Covino, "the magic and rhetoric that disappeared [during the Enlightenment]—with their emphasis on imagination, phantasy, and amplification—were *progressive* forces."[78]

Ouspensky's reading of the *Übermensch* hews close to this progressive understanding of both rhetoric and magic. Ouspensky does not want to construe magic as an object of scientific methodology. Rather, he attempts to construct evolution as this kind of progressive, magical, inventive process by which humanity (in an appropriately modernist aim) makes itself new. In an antimagical, Cartesian science, language exists as a descriptive force and, as Ouspensky argues, the "aim" of European culture is "man as he is, as he always was and always will be."[79] In such a regime of knowledge, describing in more and more perfect detail the human as it exists is the end point of all knowledge. As Ouspensky details, taking for his first case the question of scientific man and for his second case the question of magical man:

> In the first case man is taken as a completed being. Study is made of anatomical structure, his physiological and psychological functions, his present position in the world, his culture and civilisation, the possibility of better organisation of his life, his possibilities of knowledge, etc.; in all this man is taken as what he is. . . .
>
> In the second case man is taken as an uncompleted being, out of which something different should result. And the whole meaning of the existence of this being lies, in this case, in its transition into this new state. Man is regarded as a grain, as a larva, as something temporary and subject to transformation.[80]

Despite his insistence on the second case, the first case is the dominant one within Europe following the Enlightenment. For Ouspensky, evolution beyond the human is never possible so long as the process of overcoming is understood as a scientific operation.

This condition also relegates manipulators of language and the producers of written art to the peripheries of the powers that shape and create the world. This relegation, Leigh Wilson argues, is the reason for returning to magic during modernism. For the modernists,

> magic mediates between the world as it is and the world as we would like it to be as a result of our own actions. It makes it possible to act in the world in such a way that, if successful, the action would change the world, even though we know we may fail because of the world as it exists before and beyond our action.[81]

This act of world creation is also connected to the topic of invention, especially in postmodern rhetorical theory and in Fredric Jameson's understanding of Utopia (which is heavily mediated through the imagination in his discussion of the concept in *Archaeologies of Knowledge*). Through this connection to Utopian invention—to literal world-making—we can start to see the shape of transhumanism as a magical discourse. For all of its very serious commitments to science and technology, the uncomfortable aspects of transhumanism mentioned earlier in this chapter, namely the specifically evolutionary futurist ones, point not to the description of the human but toward the creation of an imagined new world, just as magic and the imagination were thought to work by modernists such as Ouspensky.

We can intensify this claim about transhumanism and magic by thinking about transhumanism as a science. If it is a science, what is its object of description? What does transhumanism seek to understand? If anything, the object most described in scientific terms by evolutionary futurist rhetoric is the future, but how can we make a science of the future? Specifically, how would a science of the future work within the confines of Karl Popper's influential theorization of science through the falsifiable, in which "the scientific status of a theory is its falsifiability, or refutability, or testability."[82] For Popper, the ability of a hypothesis to be scientific is in its ability to be tested and found to be false. Theories about the future, for all their basis in science, cannot, then, be scientific, as evidenced by the ever-receding date for the Singularity in each of Raymond Kurzweil's books on evolutionary machines.[83] The future always arrives through the passage of time, but if a prediction does not, it just signals that it may still be yet to come, not that it was false (after all, "wait and see" is not a great method for verifying experimental findings). Thus, we have to find another term for the belief system motivating transhuman action, which includes scientific research. Given that it is not descriptive of the human as it is but instead a system for imagining and then creating another world, I have shown that, following Ouspensky, transhumanism functions analogously to magical systems that seek to use collective will and language to actualize a desired future.

Magic satisfies two methodological goals for Ouspensky. On the one hand, as a methodology, magic, in contrast to science, is a communal knowledge system for manufacturing a desired future. Additionally, in the magical tradition, Ouspensky finds a blueprint for actualizing his evolutionary futurist understanding of the *Übermensch*. Synthesizing a number of esoteric topics, Ouspensky points out that the perspective necessary for overcoming the human lies, as with Loy's poetry, in cosmic consciousness, specifically in a conception of the oneness of being. However, Ouspensky complicates this unity by suggesting that "man finds superman within himself when he begins to look for him outside himself, and he can find superman outside himself when he has begun to look for him within himself."[84] The path toward *Übermensch,* then, leads through an exploration of the oscillation between self and other, while also pointing toward the kind of overcoming mandated by Bucke's theory of cosmic consciousness: the ability to grasp the self *as* other.

This double move outside and inside, for Ouspensky, points to the ancient practices of initiation, such as the Eleusinian Mysteries in ancient Greece. These practices were designed to instruct participants in an awareness of this porous boundary between self and not-self. Moreover, initiatory practices are designed to produce ecstasy in the participants, regarding which Ouspensky insists that "the normal psychic state of superman constitutes what we call ecstasy in all possible meanings of this word."[85] However, "ecstasy," associated with mystical and esoteric practice, has a number of definitions in the *Oxford English Dictionary*, suggesting how this term has mutated. The oldest *OED* definition is "the state of being 'beside oneself,' thrown into a frenzy or a stupor, with anxiety, astonishment, fear, or passion." This usage is the closest to the Greek origins of the term, from the words meaning "to be besides oneself." Later, the word came to mean "an exalted state of feeling which engrosses the mind to the exclusion of thought; rapture, transport. Now chiefly, Intense or rapturous delight." It is this denotation that Ouspensky most likely intended, but two other definitions are also suggestive: "Used by mystical writers as the technical name for the state of rapture in which the body was supposed to become incapable of sensation, while the soul was engaged in the contemplation of divine things" and "The state of trance supposed to be a concomitant of prophetic inspiration; hence, Poetic frenzy or rapture." The various meanings attached to "ecstasy" suggest a heightened state of consciousness in which individuals are in some way able to exceed the boundaries of their own consciousnesses.

In this way, the mystical rhetoric of ecstasy becomes important for Ouspensky's transhuman understanding of the *Übermensch*. If, as Ouspensky highlights in Nietzsche, the *Übermensch* is an unpredictable lightning bolt emerging from the human, the question of knowability becomes paramount. This recognition explains why transhumanism does so much rhetorical work to manufacture itself as a science of the future: the position of the prophet is not easily assimilable into a cybernetic knowledge economy. That said, ecstasy remains the best metaphor available to Ouspensky for evolving beyond the human, as it represents a state of being "so far superior to all other experiences possible to man."[86] This experience of higher evolutionary consciousness is also a rhetorical problem: "we have neither words nor means for the description of it. Men who have experienced ecstasy have often attempted to communicate to others what they have experienced . . . a description in plain words of the experiences of ecstasy presents almost insurmountable difficulties."[87] The transhuman initiate literally cannot speak of a cosmic consciousness.

This rhetorical difficulty points back to Mina Loy's experimental poetry. Her writing chafes against conventions of typography, structure, and meter, but also, and more important, against ideas of stable imagery and perspective. As Ouspensky writes, "only art, that is poetry, music, painting, architecture, can succeed in transmitting, though in a very feeble way, the real content of ecstasy."[88] From this aesthetic model, Ouspensky reiterates this initiatory mode of evolutionary consciousness by defining ecstasy "as the highest degree of emotional experience."[89] This highest emotional experience, figured as a glimpse of specifically transhuman consciousness, mirrors Loy's futurist canon: poems about love and about childbirth are what allow her to dissolve her Cartesian subjectivity into a web of intersubjective, transspecies affect that oscillates between the interior experience of pain and the ripples those experiences create through the concentric circles her being inscribes on the cosmos. As Ouspensky suggests, after a transhuman initiation, the initiate "finds a deep and strange significance in things which formerly looked self-evident and uninteresting."[90] The seemingly mundane concerns rejected by the Italian futurists, but valorized in Loy's cosmic poetics, suggest this kind of deeply symbolic emergence.

What changes from Loy to Ouspensky, however, is that Ouspensky is invested in actualizing this type of cosmic consciousness to effect a general evolutionary conversion into a transhuman being. Where Loy gives us fleeting glimpses of cosmos and the products of unnatural selection that will inherit this ontology, Ouspensky uses Nietzsche, in conversation with Darwin, to imagine practices of both self and other that can be leveraged to explore gradients of consciousness beyond the everyday. As Ouspensky writes, "the development of man towards superman cannot consist in the growth of the intellect alone. Emotional life must also evolve, in certain not easily comprehensible forms. And the chief change in man must come precisely from the evolution of emotional life."[91] Ouspensky argues that ecstasy is a glimpse of the "higher mind" toward which human evolution must be directed.

Ouspensky's "higher mind" departs from F. T. Marinetti's futurism, encapsulated in his essay "Extended Man and the Kingdom of the Machine" (1910), which projects a "nonhuman, mechanical species, built for constant speed," a species that "will quite naturally be cruel, omniscient, and warlike," as the evolutionary future for humanity. By contrast, Loy and Ouspensky offer not "more" as an evolutionary watchword but "different."[92] Loy and Ouspensky both insist on the evolution of emotion and cognition as the primary vehicles for actualizing this different consciousness. As I discuss in more detail in the next section, the chief lesson of this emotional evolution is that new forms of embodiment and the new technologies that mediate this embodiment—especially communicative technologies that extend human consciousness to a global scale—do not come without a concomitant change in consciousness as well. Unlike contemporary transhumanism, this magical evolutionary futurism emphasizes changes in consciousness at the expense of those of technology.

The Mind's Transhumanism

Having explored the modernist roots of evolutionary futurism, their similarity to the rhetorical modalities of transhumanism, and their insistence on a transhuman aesthetic and on the evolution of mind and soul as key elements in the manifestation of a new transhuman sensibility, we can reconsider Nick Bostrom's claims about Nietzsche and transhumanism that began this chapter. To reiterate, Bostrom argues that with the concept of *Übermensch,*

> What Nietzsche had in mind, however, was not technological transformation but rather a kind of soaring personal growth and cultural refinement in exceptional individuals (who he thought would have to overcome the life-sapping "slave-morality" of Christianity). Despite some surface-level similarities with the Nietzschean vision, transhumanism—with its Enlightenment roots, its emphasis on individual liberties, and its humanistic concern for the welfare of all humans (and other sentient beings)—probably has as much or more in common with Nietzsche's contemporary J. S. Mill, the English liberal thinker and utilitarian.[93]

Having seen how the question of the *Übermensch* plays out in the work of two key modernist interpreters of Nietzsche, Bostrom's terminological distinctions now appear rather confusing. Earlier in Bostrom's history of transhumanism, he supplies a definition of transhumanism as "the quest to transcend our natural confines," whether that be "socially, geographically, or mentally."[94] When elucidating this point, Bostrom is arguing that we can consider *The Epic of Gilgamesh* as an early example of the yearning for this quest. However, given that Ouspensky also returns to the idea of the mythic hero in his genealogy of the *Übermensch,* how can we unpack Bostrom's claim that Nietzsche is not related to transhumanism beyond "some surface-level similarities"? As we have seen in Mina Loy and P. D. Ouspensky, Nietzsche's quest for "soaring personal growth and cultural refinement in exceptional individuals" in the hands of those involved in Futurism and Theosophy begins to look an awful lot like the machinic overcoming Bostrom wants transhumanism to become.

Bostrom's objection to Nietzsche, with which we began the chapter, hinges on the same misreading of *Übermensch* as we found in many of the *JET* responses to Sorgner—specifically, that the term refers to individuals and not the evolution of life that comes to replace us. Although Sorgner is careful to show in his various responses that contemporary Nietzsche scholarship has recovered a communitarian vision underscoring *Übermensch,* the tension still appears to stand. Moreover, given the association between attempts to improve humanity and the specter of eugenics that often haunts contemporary transhumanism, Bostrom is strategic in disarticulating transhumanism's seeming "humanistic concern for the welfare of all humans" from Nietzsche's vaguely *unheimlich* "soaring personal growth and cultural refinement in exceptional individuals." Moreover, Bostrom *wants* to articulate transhumanism as a project whereby everyone can be saved (despite the libertarian overtones that often come to be associated with this project). Moreover, in favorably evaluating *Brave New World,* Aldous Huxley's transhuman dystopian novel, Bostrom suggests that the society Huxley warns against is specifically dystopian because it discourages "high art, individuality, knowledge of history, and romantic love."[95] It would seem that on the one hand, individuality and cultural refinement are the problem (in Bostrom's reading of Nietzsche) and the best of life to be cherished and defended (in Bostrom's reading of Huxley). Further, given that Bostrom continues to stress that transhumanists consider sacrosanct the individual's ability to choose how best to enhance his or her body, one could continue to wonder at Bostrom's dismissal, even after reading Sorgner's various defenses of a transhuman Nietzsche.

Partly, this difficulty in sorting out the refined individual from the redeemed masses results from some rhetorical slippage in the understanding of evolution. Transhumanism, as Bostrom documents it, might comprise an elite who are uniquely positioned to decide on the good and the right technologies to pursue, but, for them, this technological benevolence will benefit all through evolutionary change. In contrast, Nietzsche, as Bostrom characterizes him, imagines an evolutionary hoarding: unique individuals soar and hoard and do not attain for the evolution of all. Despite this distinction between evolution of the masses and evolution of the few, discussions of mechanisms for biological evolution tell a much more nuanced story than this hard-and-fast distinction would suggest. In Ouspensky's commentary on Nietzsche, he reminds us that "evolution, which is usually regarded as evolution of the masses, can in reality never be anything but evolution of the few," echoing this distinction between transhumanism's supposed populism and Nietzsche's ambivalence to the masses.[96] This observation reveals Ouspensky as an astute reader of Darwin. This claim about "evolution of the few" speaks to an old controversy within evolutionary theory: the question of "the units of selection."

In evolutionary biology, the unit of selection refers to the part of an organism (varying in size from individual genes to larger units such as species or ecosystems) that determines its relation to others of its species, and therefore ultimately determines its fitness. The dispute over the unit of selection fully manifested following the publication of Richard Dawkins's *The Selfish Gene* in 1976, although the controversy had been simmering since the publication of *On the Origin of Species.* Initially, in Darwin, the primary unit of selection was the individual organism, with phenotypic expression of genetic material mattering in natural selection. With the discovery of DNA and the rise of molecular biology, evolutionary biologists began to suspect, more and more, that the unit of selection was, in fact, the gene, not the organism bearing that gene. Dawkins merely condensed this thesis and focused it to a point where he could claim that, essentially, the individual organism does not matter from an evolutionary standpoint.

Dawkins would later clarify the relationship between the selfish gene and the unit of selection in the essay "God's Utility Function." In it, he argues that the unit of selection is a utility function—the term in microeconomics for strategies individuals use to maximize happiness—by which individual organisms pursue selfish genetic gains. He writes:

> Time and again, cooperative restraint is thwarted by its own internal instability. God's Utility Function seldom turns out to be the greatest good for the greatest number. God's Utility Function betrays its origins in an uncoordinated scramble for selfish gain.
>
> Humans have a rather endearing tendency to assume that welfare means group welfare, that "good" means the good of society, the future well-being of the species or even of the ecosystem.[97]

In Dawkins's theory of the selfish gene, the only traits selected for by evolution are those that allow for the maximal spread of an individual's genes. Additionally, as we can see in the quote above, the essay is an attack on utilitarianism and other forms of utopianism that seek to establish a greater good beyond the selfish gains of individuals. Moreover, Dawkins's theory of the selfish gene begins to suggest the "few" of Ouspensky's understanding of evolutionary change. For Ouspensky, only the fittest of a species can drive the evolution of consciousness toward the *Übermensch.* Similarly, in Dawkins's account of elephant seals in "God's Utility Function," he discusses a species in which "4 percent of the males accounted for 88 percent of all the copulations."[98] In other words, evolution of consciousness and evolution of the organism both seem to proceed along a winner-takes-all model in which the fittest individuals get to influence the character of the next generation.

However, does selfish gain drive the *Übermensch*? In Dawkins's model of evolution, the winner is the one who spreads the most of his (for Dawkins, anyway, it is always "his") genes through a population, thus ensuring that the fittest traits are passed on. However, Ouspensky observes, in discussing the *Übermensch,* that these evolved consciousnesses have a very bad habit of dying violently: crucifixion, stoning, burning-at-the-stake all mark the lonely and short lives he describes as being more mentally evolved. For Dawkins, this bad habit proves the biological invalidity of Utopian schemes:

> God's Utility Function, as derived from a contemplation of the nuts and bolts of natural selection, turns out to be sadly at odds with such utopian visions. To be sure, there are occasions when genes may maximize their selfish welfare at their level, by programming unselfish cooperation, or even self-sacrifice, by the organism at its level. But group welfare is always a fortuitous consequence, not a primary drive. This is the meaning of "the selfish gene."[99]

In the debate spurred by Dawkins's theory of the selfish gene, one argument against this vision developed around questions similar to those that Ouspensky seemingly offers. Specifically, the persistence of traits that do not result in *individual* reproductive gains occurs too often in biology for these traits to be the "fortuitous consequence" that Dawkins labels them. Philosopher of biology Elliott Sober jumped into this debate by discussing, broadly, the idea of altruistic genetic traits:

> An altruistic trait is one that is deleterious to the individual possessing it but advantageous for the group in which it occurs. If the organism is the exclusive unit of selection, then natural selection works *against* the evolution of altruism. If the group is sometimes a unit of selection, then natural selection sometimes *favors* altruistic traits. The units of selection problem cannot be settled by stipulative convention, because different views about the units of selection make contrary predictions about which traits evolve under natural selection. The important point is that there can be *conflicts of interest* between objects at different levels of organization: What is good for the group may not be good for the organism.[100]

As we can see, the presence of altruistic traits in animals suggests that something beyond the selfish individualism of Dawkins's model may be at work in biological evolution.

Indeed, this paradox is already present in Charles Darwin's *The Descent of Man*:

> But it may be asked, how within the limits of the same tribe did a large number of members first become endowed with these social and moral qualities, and how was the standard of excellence raised? It is extremely doubtful

> whether the offspring of the more sympathetic and benevolent parents, or of those who were the most faithful to their comrades, would be reared in greater numbers than the children of selfish and treacherous parents belonging to the same tribe. He who was ready to sacrifice his life, as many a savage has been, rather than betray his comrades, would often leave no offspring to inherit his noble nature. The bravest men, who were always willing to come to the front in war, and who freely risked their lives for others, would on an average perish in larger numbers than other men. Therefore, it hardly seems probable that the number of men gifted with such virtues, or that the standard of their excellence, could be increased through natural selection, that is, by the survival of the fittest; for we are not here speaking of one tribe being victorious over another.[101]

As we can see here, Darwin questions how common human traits could be selected for when they, in fact, work at odds with the demands of the selfish gene. Darwin then answers his own question:

> It must not be forgotten that although a high standard of morality gives but a slight or no advantage to each individual man and his children over the other men of the same tribe, yet that an increase in the number of well-endowed men and an advancement in the standard of morality will certainly give an immense advantage to one tribe over another. A tribe including many members who, from possessing in a high degree the spirit of patriotism, fidelity, obedience, courage, and sympathy, were always ready to aid one another, and to sacrifice themselves for the common good, would be victorious over most other tribes; and this would be natural selection. At all times throughout the world tribes have supplanted other tribes; and as morality is one important element in their success, the standard of morality and the number of well-endowed men will thus everywhere tend to rise and increase.[102]

In the debate sparked by the selfish gene, biologists returned to passages such as these in Darwin, armed with the language of altruism summarized by Sober, to argue that, contra Dawkins, there exists the possibility of a unit of selection larger than the gene. Curiously, it is also larger than the individual: specifically, evolutionary biologists began to consider, thanks to the study of altruistic traits, the idea that community well-being may actually be a viable unit of selection, especially in highly social animal species.

Despite Dawkins's stark assertion in "God's Utility Function," there is a long history for arguing that altruism can be an evolutionary advantage, when the unit of selection scales up from the individual to larger units such as population or species. These larger units of selection structure Ouspensky's reading of Nietzsche: the *Übermensch* acts as a signpost toward a higher order of consciousness and a higher mode of human existence. This pointing does not often benefit evolved individuals so much as it allows them to live as a means of pointing toward the future. Altruistic traits thus become hugely important to cognitive evolution and an inner transhumanism. Ouspensky, following his discussion about the evolution of the few and the evolution of the masses, suggests that "*in mankind* such an evolution can only be conscious. It is only degeneration which can proceed unconsciously in men."[103] Thinking beyond altruism, at the level of a community or even a cosmic awareness, can we not think about cosmic consciousness, as practiced by Loy and Ouspensky, as being a unit of selection for evolutionary futurism?

An obscure book by one of the chief architects of modern synthesis may provide the answer. In 1959, the biologist Bernhard Rensch published *Homo Sapiens: From Man to Demigod* in order to answer the question "what are we, we humans?" by way of the modern evolutionary synthesis Rensch helped to create with Julian Huxley and other biologists of the 1930s and 1940s.[104] Rensch suggestively proposes a three-fold approach to discussing the evolutionary past and future of human beings:

> We shall consider how the human mind emerged from earlier animal stages. With the rise and progress of cultures we shall have to sketch the outlines of man's strange and unique position, touching on certain imperfections and various dangers in the path of his future evolution. We shall see that although it springs from an entirely different root the rise of man's culture has been subject to similar laws and constraints to those that have governed the evolution of other living organisms . . . we shall then be able to attempt a cautious forecast of the future development of man's specifically human characteristics, as well as to outline a clear-cut aim—namely, a rather more purposive progressive development of man's special features and a more consistent direction of his cultural future . . . The ensuing chapters will be concerned with biological, psychological, and cultural facts.[105]

While Rensch is primarily concerned with the cultural dimension of evolution, his three folds are outlined in the final sentence of the selection: we can see that he views evolution as proceeding along three paths: informational, economic, and cognitive. This expansion of evolution's domain is important because one of the common criticisms of Dawkins lies in his lack of focus on the human. Obviously, focusing on human evolution is not the task at hand for Dawkins, but the selfish gene does not account for the singularity of human cognition, which Rensch's work does attempt to encompass. Rensch, though he does not use the language of a magical evolution, develops a sensibility similar to Bucke's cosmic consciousness in accounting for this singularity. He writes,

> The separate achievements of leading nations are fusing into a universal culture; and this culture is expanding at an immense pace, now that all sections of their populations are able to rise to higher social levels—and especially as the backward countries are making spectacular strides towards a more advanced civilization.[106]

Once again, we find a cosmic perspective operating in a discourse of evolutionary biology, this time in the work of a major figure of the modern synthesis. If we read Rensch's rhetoric carefully, the cosmic move up in scale from human to nation to globe is at work in his suggestion that industrialized nations are the units of selection driving human evolution. This kind of argument, a justification for the kind of United Nations humanism so popular at the middle of the twentieth century (and extended into the classic neoliberal apologia, *The End of History and the Last Man*, by noted critic of transhumanism Francis Fukuyama), also serves to suggest a phase shift to a broader perspective on what counts as a "unit" in a globalized human evolutionary framework.

The problem with Rensch's framework, as with many accounts of transhumanism that appeal to reason (and I would emphatically *not* lump Mina Loy and P. D. Ouspensky in this category), is the difficulty of visualizing what it would look like in a way that is not, frankly, horrifying. It is one thing to say that "countries are the unit of human evolution," and another altogether to think through its implications of nationalism and scientific racism. Saying something like, "Humans will evolutionarily merge with computers to become immortal superbeings," presents similar problems, as we saw in Hayles's critique of extropianism in the Introduction. In not considering the full implications of these claims, transhumanists often appear naive or unsubtle in their argumentation, although, considered within their own argumentative frameworks, they are just as likely walking a careful line between the rational presentation of the scientist and the metaphysical speculations of the prophet.

In a posthumously published essay titled "History of Religion and Eros," Mina Loy traces an ecstatic experience of cosmic consciousness along both of these modes of argumentation. At the end of this discussion, she reveals that "the experience of illumination was incommunicable; could only, in description be hinted at."[107] This mirrors a point Ouspensky makes in his commentary—that mystic experiences of ecstasy are seemingly incommensurate with language. This incommensurability speaks to Bostrom's conflicted terms in disconnecting the concept of the *Übermensch* from transhumanism: it is hard to say what transhumanism is not when the discourse is inherently a speculative, magical rhetoric of the future. As I mentioned above, the seeming lack of intentional ethical, philosophical, or economic outcomes in transhuman argumentation stems from a commitment to rationality and a phobia of the more magical aspects of the broader discourse of evolutionary futurism.

To this end, for all of Bostrom's desire for a new kind of human or a new, coevolved human–machine hybrid, there is a refrain of "human, all too human" in Bostrom's writing: transhumanism is about "individual liberties" and "humanistic concern for the welfare of all humans," despite also calling for an overcoming of the very category of human that underscores these humanist commitments. I attribute this contradiction partly to the basic and often unacknowledged difficulty of thinking beyond the category of human. (This is why so much science fiction never experiments with basic human categories.) However, as Ouspensky reminds us, the "superman cannot be simply a 'great business man' or a 'great conqueror' or a 'great statesman' or a 'great scientist.' He must inevitably be either a magician or a saint."[108] These categories all illuminate one necessity running through the history of transhumanism: if the basic axioms of evolutionary futurism are taken *very* seriously, the conversation surrounding transhumanism cannot be one invested in perfecting existing human forms but instead one of displacing those forms in favor of something else.

This displacement may not always have been intentional, but it is always embedded in transhuman argumentation. Many extropian transhumanists believe that nothing will change when they upload their consciousnesses into computers, but magical evolutionary futurism simply asks: How can this be? Transhumanists are often, argumentatively, in a double bind: on the one hand, they appear dangerous by being optimistic about normal humans living normal lives as computer programs, on the other they appear unhinged if they make grand claims of cybernetic, bodiless supermen.

In Loy's account of mysticism, religions, and the erotic in "History of Religion and Eros," this difficulty of communicating *any* vision of ecstatic becoming (not only one mediated through computers) becomes not just a problem of communication but a problem in which to retransmit a message is to degrade the clarity of the message. She writes that, "to the earliest disciples, the mystic system proved often communicable. But all systems of spiritual exercise come to a crisis incurring decrease of comprehension."[109] These diminishing returns signal the inevitable collapse of mysticism into gibberish, rote practices of faith without a core of philosophy behind these now-empty gestures. Loy's powerful move beyond this pessimism, however, is to assert that "the snap-back of human consciousness from the take-off of inspiration: the *stretch* of consciousness into the imperceptible universe still depend[s], to some degree, on the contemporary stage of evolution in the *concrete world*."[110] In other words, Loy recognizes that these original visions of ecstasy beyond the human have to be tied to an evolutionary uptake in consciousness, as seeing the concrete world differently is key to unlocking the imperceptible.

It is not enough, then, to posit a future humanity that has merged with computers or that lives immensely long lives in robot bodies or that begins building robot arms or that takes a hundred supplements a day or that follows whatever other technological impulse or commits to whatever other tropes one might associate with transhumanism. The inner transhumanism of this mystical, modernist evolutionary futurism reminds us that these visions of an enhanced, perhaps revolutionary, posthumanity further demand an expansion in our ability to comprehend that future—and to comprehend it as alien. That transhumanism is often ominous or laughable in its seeming naive faith in the future, yet its comfort with violent technological disruption of basic human existence, as Ouspensky and Loy show us, results from the basic incommensurability between the ecstatic content of the transhuman vision and language's ability to transmit that vision. Perhaps, then, the answer is not more data or more gadgets but a more mystic transhumanism. Otherwise, we are left to conclude that contemporary transhumanism attempts to build a mysticism without the mystical.

2 ASTOUNDING TRANSHUMANISM!

Evolutionary Supermen and the Golden Age of Science Fiction

Answering the editor's topic question for the September 2004 issue of *Foreign Policy,* "What Is the World's Most Dangerous Idea?" Francis Fukuyama described transhumanism—positioned as the dangerous idea between "Spreading Democracy" and "Religious Intolerance"—as "a strange libertarian movement" whose members "want nothing less than to liberate the human race from its biological constraints."[1] This "odd cult," where some members "freeze themselves cryogenically in hopes of being revived in a future age," stands to create a divide between evolutionary haves and evolutionary have-nots that destabilizes the very idea of a universal human that, as Fukuyama argues, underscores our basic notions of rights and ethics.[2] At the same time—in the early years of the War on Terror—that Fukuyama proselytizes about the imminent and massive threat to democracy posed by transhumanism (which he says pales in comparison to the more immediate threats of terror and amok government surveillance), he dismisses transhumanism as "nothing more than science fiction taken too seriously."[3] Fukuyama attempts to dismiss transhumanism by aligning the movement with the archetypally nerdy image of the science fiction fan—such as Comic Book Guy on *The Simpsons* or any of the stereotypical male leads on *Big Bang Theory.* And yet he takes very seriously the idea that transhumanism's promise of an evolutionary future poses a basic threat to human rights. While he attempts to align transhumanists with the socially awkward and arrogant stereotypes of the nerd, he also warns that transhumanists "deface humanity with their genetic bulldozers and psychotropic shopping malls" and that we increasingly live in a world in which we "nibble at biotechnology's tempting offerings without realizing that they come at a frightful moral cost."[4] In this essay, Fukuyama argues that transhumanism takes science fiction too seriously; at the same time he is warning us that we all must start taking science fiction too seriously, too.

In this chapter, I take up two issues with this assertion. First, I consider how Fukuyama is able to be simultaneously so dismissive and so concerned about transhumanism's version of evolutionary futurism. Moreover, I situate Fukuyama as a critic of the transhuman movement and outline how his position, labeled "bioconservative" in the transhumanist literature, has played out in the further formulation of a transhuman ethics. Second, and constituting the bulk of this chapter, I take issue with the assertion that taking science fiction seriously is a problem in the first place. To this end, I consider the role evolutionary futurist rhetoric, in the figure of the genetic superman, played in the formation of modern science fiction. I explore the culture of the Superman Boom—a period that coincided with the dawn of science fiction's Golden Age in the 1930s—and how the utopian responses to A. E. van Vogt's novel *Slan* played a role in birthing modern fan culture through a specifically transhuman rhetoric. From this milieu, I argue that evolutionary futurism was an integral rhetorical ingredient in the emergence of science fiction as a utopian mode for critiquing technology, bioethics, and the future evolutionary dimensions of humanity.

The Specter of the Bioconservative

Francis Fukuyama's dismissal of transhumanism as "science fiction taken too seriously" in the same essay in which he hails it as "the most dangerous idea in the world" is embedded in a much longer criticism of transhumanism, specifically of its agenda of enhancing the human body through technological means—genetic engineering, life extension, cybernetic prostheses. For Fukuyama, these technologies of enhancement threaten to divide humans into different species with different classes of rights. In his thought, human beings have equal, basic rights because "underlying this idea of the equality of rights is the belief that we all possess a human essence that dwarfs manifest differences in skin color, beauty, and even intelligence."[5] Through enhancement, this precious bodily essence is threatened with fragmentation, in which a genetic underclass and overclass exist within an unequal distribution of political rights. If this sounds remarkably like contemporary geopolitics, or colonialism or the Jim Crow American South or Rhodesia or Raj-era India, you are probably right, but it is important to note the thesis of Fukuyama's *The End of History and the*

Last Man (1992), which he summarizes in the *Foreign Policy* piece: "slowly and painfully, advanced societies have realized that simply being human entitles a person to political and legal equality."[6] I will not rehearse the myriad arguments against this core of Fukuyama's theory of geopolitics, but it is important to note that the idea of an eventual arrival at universal human rights underscores his objections to transhumanism. He regards transhumanism as a fundamental barrier to our ever attaining universal human dignity based on a shared species-being.

Fukuyama's piece in *Foreign Policy* extends the argument from his book *Our Posthuman Future* (2002). The publication of this book played a major role in a watershed moment in the history of transhumanism, as James Hughes makes clear in his introduction to a special issue of *The Journal of Evolution & Technology (JET)*:

> In 2001, conservative philosopher Leon Kass, an outspoken opponent of in-vitro fertilization, stem cell research, and life extension, had been appointed by George Bush to head the U.S. President's Council on Bioethics (PCB). Kass subsequently appointed a number of conservative intellectuals to that body, including Francis Fukuyama and Charles Krauthammer. Under Kass and Fukuyama's direction the PCB's first order of business was the supposed threat to humanity from human enhancement technologies.
>
> In 2002, Francis Fukuyama published *Our Posthuman Future*, which argued for global treaties to restrict enhancement technologies that he argued threatened the foundation of human rights. Also in 2002, Leon Kass published *Life, Liberty, and the Defense of Dignity*, which argued that life extension and other enhancements were "dehumanizing." In 2003, the PCB published its own critique of human enhancement, *Beyond Therapy*, reflecting many of Kass and Fukuyama's concerns about "better children," "ageless bodies," and "happy souls." The journal *New Atlantis* was also created in 2003 at the conservative Washington thinktank the Ethics and Public Policy Center to work closely with Kass and the PCB to promote this new conservative critique of enhancement technologies.[7]

This collection of events, along with the founding of a laundry list of advocacy organizations in 2003, comprises what Hughes notes as the emergence of "a diverse coalition of 'bioconservative' groups on the left, right, and center."[8]

"Bioconservative" first emerges as a designation around 2004 in the published work of Nick Bostrom and represents what Bostrom calls "transhumanism's opposite."[9] As Bostrom illuminates, this new strand of cultural conservatism represents an interesting mutation in the hard-and-fast political boundaries between liberal and conservative in Western politics. Bostrom, assuming that conservatives, who nominally value self-determination and freedom from government, would align with transhumanism's emphasis on personal liberty, notes that, on the contrary,

> they have gravitated towards transhumanism's opposite, bioconservatism, which opposes the use of technology to expand human capacities or to modify aspects of our biological nature. People drawn to bioconservatism come from groups that traditionally have had little in common. Right-wing religious conservatives and left-wing environmentalists and anti-globalists have found common causes, for example in their opposition to the genetic modification of humans.[10]

As Bostrom continues, "nowadays [bioconservatism] commonly emanates in calls for national or international bans on various human enhancement technologies."[11] This motley alliance of environmentalists, religious groups, anti-globalization activists, and pro-globalization philosophers such as Fukuyama stands united in agreement with the title of Bill McKibben's bioconservative classic, *Enough: Staying Human in an Engineered Age* (2003).

One of the most prominent examples of this newly emergent bioconservative agenda was the puzzling, perhaps even absurd, reference to "human–animal hybrids" in George W. Bush's 2006 State of the Union Address:

> Tonight I ask you to pass legislation to prohibit the most egregious abuses of medical research: human cloning in all its forms; creating or implanting embryos for experiments; *creating human–animal hybrids*; and buying, selling, or patenting human embryos. (Emphasis added)

Obviously a result of Kass and Fukuyama's involvement in science policy in the Bush White House, this reference, sandwiched between discussions of health care initiatives and congressional ethics reform, seemed strange at the time, but is part of the bioconservative agenda against transhuman technologies that became a centerpiece of science policy in the George W. Bush White House. However, it also highlights two contradictory yet coexisting points about transhumanism. On the one hand, it shows how difficult it is to reference these issues without sounding like someone who has watched too much *Star Trek*. On the other, it shows the seriousness with which policy makers are coming to view evolutionary futurism.

This emergence of a bioconservative response to transhumanism represents a major turning point in contemporary transhumanism. As I argued in chapter 1, evolutionary futurism is a speculative discourse more akin to a magic than a science. In this moment of acknowledgment, a call by the president of the United States to take seriously the idea of legislating against some of the more gonzo claims made by transhumanists throughout the twentieth century, transhumanism went from being a purely speculative utopian ideology imagining the ins and outs of an evolutionary futurist posthumanity to becoming a stakeholder in a political struggle over the bioethics of humanity's future. As Nick Bostrom summarizes,

> Both agree that we face a realistic prospect that technology could be used to substantially transform the human condition in this century. Both agree that this imposes an obligation on the current generation to think hard about the practical and ethical implications. Both are concerned with medical risks of side-effects, of course, although bioconservatives are more worried that the technology might succeed than that it might fail. Both camps agree that technology in general and medicine in particular have a legitimate role to play.[12]

In other words, suddenly the *topoi* of transhumanism were no longer the speculative content of *a* future: they were the political content of *the* future. We no longer needed science fiction to tell us about this future; now, we need bioethics to legislate the limits to which such technology can be pushed and applied to our bodies. The switch from utopia to politics signaled by Bush's call meant that the transhumanists were no longer alone: *everyone* was taking science fiction seriously.

Science Fiction Conquers the World!

This seriousness was not always the case, obviously. Philip K. Dick, in "How to Build a Universe That Doesn't Fall Apart Two Days Later," his classic essay on world-building and the craft of science fiction, explains the plight of the science fiction writer by recalling how "friends would say to me, 'But are you writing anything serious?' meaning 'Are you writing anything other than science fiction?'"[13] From this question of seriousness and speculation, Dick unlocks some of his gnostic understandings of his 2-3-74 experience that also inspired his "Exegesis,"[14] but before diving into the more esoteric aspects of his argument, he begins by questioning just what science fiction writers are actually good at. Dick laments that "we can't talk about science, because our knowledge of it is limited and unofficial, and usually our fiction is dreadful."[15] However, he does suggest that science fiction is good at building worlds and making them stick in the mind. More importantly, he suggests that, for him, a life of writing science fiction has been motivated by "two basic topics which fascinate me . . . 'What is reality?' and 'What constitutes the authentic human being?'"[16]

For Dick, attempts to answer these questions, touched on in "How to Build a Universe" and explored extensively in the eight thousand handwritten pages of his "Exegesis," are the serious content of science fiction. At the same time, however, this seriousness is problematized in Dick's reception as both a serious novelist and a serious mystic. As Dick biographer Lawrence Sutin notes in the introduction to *The Shifting Realities of Philip K. Dick* (the collection containing "How to Build a Universe"),

> To this day one finds, in SF critical circles, sharp resistance to the notion that Dick's ideas—divorced from the immediate entertainment context of his fiction—could possibly be worthy of serious consideration. It is as if, for these critics, to declare that certain of Dick's ideas make serious sense is to diminish his importance as the ultimate "mad" SF genius—a patronizing role assigned him by these selfsame critics.[17]

In SF circles, many critics tolerate Dick as a "'mad' SF genius" as a means of transforming the serious mystical content of his work into a series of eccentric affectations. Taking him seriously as a mystic and a postmodern philosopher would threaten the boundaries of science fiction as an autonomous space for outsiders to escape mainstream persecution. Seriousness thus becomes a threat to the model of the isolated and misunderstood genius that animates much of science fiction.

As Philip K. Dick explains in "How to Build a Universe," however, the rhetorical problem of seriousness haunts science fiction in a number of ways. Beyond "are you writing anything serious?" he follows through the logic of some possible answers to his "two basic topics," concluding with the observation that "I offer this merely to show that as soon as you begin to ask what is ultimately real, you right away begin to talk nonsense."[18] The cacophony of ideas circulating in Dick's "Exegesis" certainly confirms this nonsensicality. As the editors of the 2011 publication of extensive portions of the manuscript observe, "the topics [covered in "Exegesis"]—apart from suffering, pity, the nature of the universe, and the essence of tragedy—include three-eyed aliens; robots made of DNA; ancient and suppressed

Christian cults that in their essential beliefs forecasted the deep truths of Marxist theory," etc.[19] So, in other words, postmodernism.

But in all seriousness, Dick's tendency to address a myriad of topics, and his lack of commitment to the logical rigors of philosophy or the whimsical rigors of science fiction, suggest the true problem of taking science fiction seriously: Do we have to take the aliens with the angst? Or, more problematically, as Dick asks about the truth content of science fiction in "How to Build a Universe," the question is "not, Does the author or producer believe it, but—Is it true?"[20] He speculates:

> Speaking for myself, I do not know how much of my writing is true, or which parts (if any) are true. This is a potentially lethal situation. We have fiction mimicking truth, and truth mimicking fiction. We have a dangerous overlap, a dangerous blur. And in all probability it is not deliberate. In fact, that is part of the problem. You cannot legislate an author into correctly labeling his product, like a can of pudding whose ingredients are listed on the label. . . . [Y]ou cannot compel him to declare what part is true and what isn't if he himself does not know.[21]

Dick's experiences after being contacted by what may have been an artificial intelligence from the future—or the noösphere, or an acid flashback, or God—intensified his belief both in the illusory nature of reality and in the rightness of Richard Bucke's theory of cosmic consciousness: that there is a truth content to science fiction, but it operates in unusual ways. In some of the more bizarre passages of "How to Build a Universe," Dick highlights ways in which his novels have anticipated both his own personal future and the content of Biblical prophecy. These, for him, represent truth vectors in his fiction. Moreover, the vision of the Logos he develops from these revelations—"both that which thought, and the thing which it thought"—suggests an increasingly blurry interchange between science fiction and reality, an interchange in which it was unclear what is true and what is false. To reorder my earlier question: What if the angst is illusion and the aliens are, in fact, true?[22]

Whatever truth there may be to Dick's claims about "the" "truth" of science fiction (a truth I would ally with the rhetoric of seriousness animating this chapter), reading *The Exegesis of Philip K. Dick,* one is struck by the sense that Dick was convinced humanity was moving, evolutionarily, toward a revelation of the illusory reality projected by Logos and, from there, into a cosmic consciousness beyond this veil. Writing about Terence L. McKenna and Dennis J. McKenna's book *The Invisible Landscape* (1975) (itself an attempt to make sense of a series of intense psilocybin experiences), Dick muses that an "acceleration of acceleration is what took place in my brain (and hence world) in the first stages of 3-74."[23] Whether through the accumulation of information in an emerging noösphere or humanity's completion or through the completion of some kind of species-level, Christ-like penance, Dick's "Exegesis," is suffused with a sense that the truth of his fiction was extruding into everyday life.[24]

To turn from taking Dick literally to considering him metaphorically, I now want to consider a common trope in popular rhetorics of technology in the last twenty years: that of our increasingly science fictional reality. From the proliferation of digital devices that resemble communicators from *Star Trek* to the literalization of formerly hypothetical moral questions of using robots to fight our wars for us, our reality increasingly resembles our science fiction. As SF author Ken MacLeod quipped on Twitter, "I recharged my cigarette, blocked a sex robot from my timeline, and followed an astronaut. Then I wrote a science fiction story."[25] Macleod's joking aside, his tweet points to a growing sense that reality has somehow caught up with the classic rhetorical moves of science fiction.

However, despite the failings mentioned by Dick—namely, their limited knowledge of science and their embarrassing inability to articulate their ideas in an academically acceptable fashion—SF writers increasingly find themselves not kooky enough. Consider, for instance, Dick's meeting Fredric Jameson, Darko Suvin, and some of the other founders of academic science fiction studies. Pamela Jackson and Jonathan Lethem note in their introduction to *The Exegesis of Philip K. Dick,* "An interesting Exegesis subplot consists of Dick's reactions to meeting some of his earliest admirers in academia, whom he refers to as 'The Marxists' and who were clearly perplexed by his metaphysical preoccupations. 'I proved to be an idiot savant,' he writes, 'much to their disgust.'"[26] The fact that this once-shocking observation, "reality is becoming science fictional," is becoming banal is interesting, as this commonplace is actually much older than recent acceptance might suggest. In H. Bruce Franklin's *Robert A. Heinlein: America as Science Fiction* (1980), we find perhaps the most complete statement of this idea of interchange between science fiction and reality:

> The phenomenon of Robert A. Heinlein expresses, among other things, the extraordinary quality of everyday experience of our century. Heinlein is certainly our most popular author of science fiction, easily the most controversial, and perhaps the most influential. And science fiction has moved

> inexorably toward the center of American culture, shaping our imagination (more than any of us would like to admit) through movies, novels, television, comic books, simulation games, language, economic plans and investment programs, scientific research and pseudo-scientific cults, spaceships real and imaginary.[27]

In Franklin's account, science fiction exists as an unconscious entanglement with the American twentieth century itself. On the one hand, science fiction documented the meteoric explosion of global and planetary technologies that rendered it possible to imagine a galactic humanity. On the other, consumption of science fiction by engineers and policy makers led to the construction of technologies to actualize the ideas presented in science fiction. This feedback loop is not a new phenomenon, although, as Dick suggests, perhaps we are living through an acceleration of that feedback loop—an acceleration that demands taking things seriously. Alternately, as Dick's confusion with the 2-3-74 experiences suggests, perhaps we are entering a version of the SF–reality feedback loop that fundamentally destabilizes truth.

However, to consider the early moments of this feedback loop, we can turn to the 1942 murder mystery *Rocket to the Morgue* by Anthony Boucher (who wrote mysteries under the pseudonym H. H. Holmes). Boucher mentored a young Philip K. Dick and was an important editor and social fixture in the Southern California community of SF writers immediately before World War II (he was also the first to translate Jorge Luis Borges into English). Boucher's mystery novel offers a fascinating window into the early history of American science fiction, as the plot includes thinly fictionalized versions of his friends, including Robert Heinlein, L. Ron Hubbard, and Henry Kuttner as suspects, and C. L. Moore, Edmund Hamilton, John W. Campbell, and Forrest J. Ackerman as supporting characters. (The character based on L. Ron Hubbard, interestingly, is the murderer.) The novel also fictionalizes the early solid-fuel experiments at Caltech of rocket pioneer and occultist Jack Parsons, who, before his early death in an engine explosion, shared friends with Boucher and was a member of the California branch of Aleister Crowley's Ordo Templi Orientis.[28] In the novel, before a public test of one of these early rocket engines, Robert A. Heinlein's fictional analogue, Austin Carter, delivers the following speech:

> There's just one point about rockets I'd like to venture on my own before we start the demonstration. I don't know if Hugo agrees with me on this; he probably hasn't even bothered himself about it. But it's this: That the rocket

> carries in its zooming path the hopes of all men of good will. By leaving the planet, man may become worthy of his dominion over it, and *attain dominion over himself.* The realization that there is something beyond this earth, if only a purely physical sense, may unite this earth, may change men from a horde of wretchedly warring clans to a noble union of mankind.
>
> I may be deluded in my hopes. The discovery of new worlds may be as futile as the discovery of the New World. It may mean only further imperial wars of conquest, new chapters in the cruel exploitation of subject native races. But it may mean new unity, new vigor, new humanity, and the realization at last of all that is best in mankind. I hope so anyway.[29]

In Boucher's transposition of Heinlein's words, we find that the rocket was itself a symbol of an evolutionary leap in human consciousness, representing the opportunity to rethink what it means to be human and the possibility to reshape what we think of as humanity. While depicted in a thinly fictionalized variation of the SF community organized around Heinlein's Mañana Literary Society, these ideas can also be seen in a number of his novels during the early Golden Age period, especially in Heinlein's 1956 *Double Star.* For Heinlein, the rocket represents a chance to make a break from the past and create a new model of human existence. In many ways, this model of the rocket positions it as a potent transhuman symbol. In very different ways, both William S. Burroughs and Timothy Leary take up this symbolic function during the 1960s, as the idea of the rocket moves from the proper confines of science fiction tentatively into more mainstream culture. Moreover, Boucher's depiction of Heinlein's speculation suggests a further dimension to Dick's claim about "doing serious work." We might expect a mad genius such as Dick—one of the very few SF authors to attain any kind of mainstream literary respect—to be grappling with serious philosophical questions, but Heinlein's engagement with the rocket and evolutionary futurism suggests that philosophical questions and experimental humanisms are at the core of science fiction.

In sharp contrast, however, stands SF author Thomas M. Disch's vitriolic *The Dreams Our Stuff Is Made Of.* A denunciation both of Disch's own career as an SF writer and of many of his peers, Disch's history of science fiction and its influences on reality includes such gems as his description of SF fans as "nerdy teenagers and those older readers whose taste in reading was dictated by the nerd within."[30] Relevant to Boucher's depiction of Heinlein on the rocket, Disch suggests that the increasing ubiquity of SF tropes in everyday life is only a symptom of our increasing infantilization and the decay of our culture. For Disch,

> the golden age of science fiction is twelve, [and] it follows that SF writers will be successful in proportion as they can maintain the clarity and innocence of wise children. Writers as diverse as Ray Bradbury, Harlan Ellison, Anne McCaffrey, Piers Anthony, and Orson Scott Card all owe a good part of their popularity to their Peter Pannishness. Characteristically, their stories do not pay much heed to those matters of family and career that are the usual concern of mature, responsible adults and the mature, responsible novelists who write for them, like John Updike and Anne Tyler.[31]

To this end, Disch suggests that infantile themes and childish desires are all that science fiction can provide:

> In the 1920s and 1930s, when American SF was aborning, its menu of future wonders was a national letter to Santa Claus listing the toys that boys like best—invincible weapons and impressive means of transportation. When the future began to arrive, in the '50s and '60s—that is, when the dreams of the SF magazines began to be translated into the physical realities of the mature consumer culture by a generation of designers and engineers who'd come of age in the Pulp SF era—cars were streamlined to resemble rocket ships. In fact, the car was revealed as the secret meaning of the rocket ship, a symbol, at gut level, of absolute physical autonomy.[32]

Unlike the evolutionary futurist seriousness with which Heinlein treats the rocket, for Disch, SF narrative tropes are merely symbols trying to find their true desire—in the case of the rocket, "absolute physical autonomy"—and are waiting to be satisfied by the ever-ready forces of free-market capitalism. For Disch, in fact, "the complex equation of car and rocket ship epitomizes the relationship between SF and the surrounding culture."[33] This mediation by capital is key to the pathways between SF and reality in Disch's account. In reevaluating his life's work, Disch concludes that science fiction is only a tool to market an infantile and id-driven capitalist culture of accumulation to susceptible children.

In sharp contrast to Disch's formulation of infantile desire, capitalist accumulation, and the decline of civilization, Andrew Ross's essay on science fiction in 1930s-era progressivism, "Getting Out of the Gernsback Continuum," suggests that science fiction during the same period Disch is analyzing cultivated serious goals and seriously contributed to mainstream conversations about science and technology. Ross's title alludes to the famous William Gibson story, "The Gernsback Continuum"—a contemporary-set tale of a photographer capturing images of the vanishing "Raygun Gothic" landscape of the American West. While hallucinating superimpositions of a Gernsbackian future over the dreary and broken-down reality of the 1980s, the photographer realizes something profound about 1930s futurism: the naive account of techno-progressivism displayed in Gernsback's writing—the "national letter to Santa Claus" of Disch's history—allowed for the cooptation and destruction of a kind of futuristic optimism at the hands of multinational capital. In Gibson's story, the 1930s are situated as a failure to anticipate consequences of Utopian scheming:

> The Thirties dreamed white marble and slipstream chrome, immortal crystal and burnished bronze, but the rockets on the covers of the Gernsback pulps had fallen on London in the dead of night, screaming. After the war, everyone had a car—no wings for it—and the promised superhighway to drive it down, so that the sky itself darkened, and the fumes ate the marble and pitted the miracle crystal.[34]

Gibson's account of the 1930s from the perspective of the jaded 1980s cyberpunk stands as the dominant reading of this period, a reading Ross seeks to overturn.

For Ross, focusing on the figure of Hugo Gernsback—the editor famous for launching a number of early SF pulp magazines and organizing the first network of authors and fans of "scientifiction"—affords a more nuanced account of the struggle between utopian impulses and the infantile and tawdry future Disch projects. Ross's goal is not to rehabilitate Gernsback's reputation as a writer similar to the critical reevaluation of Philip K. Dick following his death; rather, Ross attempts to break from a "linear" conception of science fiction's development in which lesser-quality work is replaced by work of a higher quality.[35] Despite this goal, he concedes that the model of science fiction Gernsback championed was, in fact, naive in many cases; however, Ross suggests that decoupling science fiction from notions of quality focuses our attention on a contemporaneous "story, naive or not, that SF also tells about the place of science and technology in society."[36] Through reading a period's speculative imaginings, Ross argues, we can recover a kind of technological unconscious: a historical moment's specifically enmeshed tangle of technological change and utopian imagination.

In looking to Gernsback and early Golden Age science fiction, Ross uncovers an organized political agenda he calls "critical technocracy." In uncovering this agenda, Ross argues that this period of SF history reveals an entirely other form of progressivism than the dominant model of an agrarian populism embodied in Depression-era figures such as Woody Guthrie and Huey Long. Ross captures an era of "three decades of progressive thinking about technology's capacity to weld the future and progress together into one social shape."[37] Ross concedes Disch's understanding of the intervention of capitalism into this utopian speculation, however. Also, in a passage unrelated to Disch's discussion above, Ross considers the interchange between rockets and automobiles, suggesting that both the general utopian vision of a critical technocracy and the more specific vision of an Art Deco futurism critiqued in Gibson's tale "endowed the likes of GM with powers that soon came to preside without ecological foresight over the nuclear militarization of aerospace and the carbon-intensive automobilization of ground space."[38]

Despite this failure, Ross recuperates this period of science fiction by suggesting that "once it has abolished utopias by announcing the end of ideology, corporate technocracy has to deliver what it promises—incremental raises in consumer gratification—or it is found wanting."[39] However, the risky experiments in creating a new "social shape" for the future signal that

> SF culture is not part of that risky game. Its futures provide ample room for alternative forms of gratification. Even in those early years, when SF most embodied the technocratic spirit, there was a close link to what I have described as critical technocracy, an attempt, in its heyday at least, to change the rules of the game that have governed GM's idea of technological progress.[40]

Thus, despite influencing reality and potentially shaping the course of technological development, science fiction is not beholden to the same strictures as corporate or scientific notions of progress. No one would accuse Robert A. Heinlein of not having delivered us rocket ships.

In this way, the question of taking science fiction too seriously, at this moment, appears to mean taking seriously the idea that technological change can be determined in ways that signal lines of flight from "corporate technocracy" and toward other "social shapes." These social shapes are also the domain of contemporary transhumanism. In the previous chapter, we saw that transhumanism is a mode of magical, rather than scientific, rhetoric: the creation of a new world through a shared and focused intent. Ross's account of science fiction's critical technocracy suggests a similar function for science fiction. The literature, as Ross suggests, is not simply the tool for creating an infantilized reality that Disch makes it out to be. In Ross's account, the corporate or scientific form will always fail to satisfy the promise of science fiction, thus keeping the entire engine of desire fueled for something else. We are beginning to see that the question of taking science fiction too seriously is more akin to acknowledging the entangling of truth and fantasy that Dick suggests in his nonfiction writing on his 2-3-74 experience. To explain his concept of the "fake fake," an increasing by-product of television's ability to market fake realities to fake people, Dick offers the following example,

> in Disneyland there are fake birds worked by electric motors which emit caws and shrieks as you pass by them. Suppose some night all of us sneaked into the park with real birds and substituted them for the artificial ones. Imagine the horror the Disneyland officials would feel when they discovered the cruel hoax. Real birds! And perhaps someday even real hippos and lions. Consternation. The park being cunningly transmuted from the unreal to the real, by sinister forces . . . They would have to close down.[41]

Perhaps making the fake birds real is the true function of this creational mode, whether called magic or science fiction or transhumanism. Science fiction's true seriousness is through the logic of the prank: as SF writer Bruce Sterling has suggested, "if poets are the unacknowledged legislators of the world, science fiction writers are its court jesters."[42] This conforms to Ross's analysis, in which science fiction is not beholden to *delivering* the future it imagines. Science fiction, instead, animates our technical objects, allowing them to speak to us about what new modalities of human potential—and what new threats—they unlock. The question of taking science fiction seriously, then, starts to resemble taking seriously the fundamental axiom of evolutionary futurist rhetoric: that our technological adjuncts play an increasing role in our evolution as a species.

Fans Are Slans!

When Robert Heinlein's analogue in *Rocket to the Grave* claims that "by leaving the planet, man may become worthy of his dominion over it, and *attain dominion over himself*," he is outlining the mutations of science fiction's Utopian project from the Gernsback period to the Golden Age. Where the critical technocracy of the Gernsback period sought to popularize science through the figure of the citizen-inventor, the failure of that project, crystallized in GM's "Futurama" at the New York World's Fair, suggested that the human itself, and not access to or knowledge of technological achievement, was a central barrier to the techno-utopia imagined by science fiction. As a result of this observation, as Andrew Ross mentions, Gernsback's chief competitor "Orlin Tremaine, then the editor of *Astounding Stories,* introduced the more metaphysical 'thought-variant' story. Campbell, Tremaine's maverick successor, encouraged his famous stable of writers to try more speculative, psychohistorical, and even sociological treatments. In the fifties, Campbell allowed his writers to investigate the fields of psi, dianetics, and parapsychology."[43] The "thought-variant" story mentioned by Ross may be more immediately recognizable as Isaac Asimov's "social science fiction" story, in which political organization and mental development replace technological hype as the primary imaginative, generative work of science fiction during the Golden Age.

The first attempt at working through this thought-variant story results in a period in SF history I call the Superman Boom. During the late 1930s and early 1940s, the pages of science fiction magazines, especially Campbell's *Astounding,* were littered with tales of genetically evolved supermen. These supermen were usually smarter and more physically agile than their genetically normal readers. Additionally, they sometimes were aided by elite mental and physical conditioning. The classic example of this subgenre is A. E. van Vogt's novel *Slan,* which details the exploits of a persecuted minority of genetically evolved humans on a future Earth.

The serialization of *Slan,* van Vogt's first novel, starting in 1939 in the pages of *Astounding Science Fiction,* established the narrative and tropological patterns of the Superman Boom. The novel was also the first SF novel that a fan *had* to read to be taken seriously as a serious SF fan. The novel's status as a sensation was only further enforced when it became the first SF novel to be published in hardback, inaugurating a number of publishing firsts for van Vogt.[44] The novel, concerned equally with embarrassingly naive technological wish fulfillment and serious meditations on urbanity and evolutionary change, marked the clear beginning of the Golden Age of science fiction as something entirely new and different. Additionally, as we shall see, its author and its many readers took the themes of *Slan* very seriously, producing a wide range of Utopian, evolutionary futurist projects that differently blur the boundaries between fiction and reality, while also documenting the early days of transhumanism in the United States.

The novel follows a member of a new human-derived race of persecuted genetic superbeings, precocious teenager Jommy Cross, after the police murder his mother. In the future projected by van Vogt, scientists have bred genetically superior humans, named "slans" as a shortening of the name of their supposed creator, Samuel Lann. These slans are persecuted for their superior abilities: they are psychic and have superior mental and physical characteristics. As Jommy flees from the relative safety of the suburb in which he and his mother had been hiding, the novel charts both Jommy's individual maturation and a more broad, growing awareness of the need, on the part of all slans, to organize against their persecutors. In this section, in addition to documenting the form of this early evolutionary futurist subgenre of science fiction, I show how this superman narrative specifically fed back into reality and created some early forms of transhumanist organization among SF fans in the 1930s and 1940s.

Psionic Systems of the Supermen

In terms both of influence during the period and sheer number produced, A. E. van Vogt was John W. Campbell's most successful author of superman stories. Not only an author singularly dedicated to this moment in SF history, van Vogt was involved in several real-world attempts to systematize transhumanism's unlocking of latent genetic potential in humanity. Van Vogt, along with a number of other fans and professional SF writers (including Robert Heinlein), became interested in Alfred Korzybski's General Semantics, a system of non-Aristotelian logic. Korzybski, whose *Science and Sanity* is the source of the famous phrase "the map is not the territory," taught a system of linguistics and psychology in which the direct, unconscious association between word and thing is removed through the systematic removal of "to be" verbs from speech and thought. For Korzybski, this removal creates the possibility of higher consciousness through the unlocking of meta-cognition, the same thinking about thinking so important to Richard Bucke's cosmic consciousness. Whereas Bucke takes the more mystic cast that dominated chapter 1, Korzybski is practical and explains his practice in the language of logic, mathemetics, and, perhaps most unexpectedly, the early self-help rhetoric that promised happiness and a system for developing radical human potential. Van Vogt's 1949 novel, *The World of Null-A,* channeled the enthusiasm he and his friends felt for Korzybski's system and adapted it to the generic form of the superman narrative championed by Campbell.

In this milieu, Campbell first published L. Ron Hubbard's *Dianetics* in the May 1950 issue of *Astounding Science Fiction*. Like General Semantics, Hubbard's Dianetics promised to convert its users into higher forms of consciousness through a variety of pseudo-cybernetic techniques. Campbell wrote an afterword on the science of Hubbard's claims for the first hardback edition of *Dianetics* (excised from later editions, following the Dianetics Group's conversion to the Church of Scientology), and van Vogt served as the Dianetics Group's first treasurer, before leaving when the group transitioned into the modern Church of Scientology. Like General Semantics, the interest in Dianetics was partly motivated, for Campbell and van Vogt, by enthusiasm to create real-world versions of the supermen they were producing in fiction. Unlike his superman narrative about General Semantics, van Vogt never published the novel about Dianetics he alludes to in his autobiography.[45] In any case, Campbell and his star author shared enthusiasm for a kind of systematic approach to parapsychology that prefigures the Human Potential Movement of the countercultural 1960s. Both were heavily involved in a variety of schemes to unlock human potential, all inspired by the evolutionary futurist imperatives of the superman fiction they were involved in producing.

In this climate of real-world, proto-transhuman experimentation in parapsychology, Campbell wrote numerous editorials advocating for the study of paranormal phenomena. In fact, this desire for such a precise approach to the mind and its expansion was an organizing force during the Golden Age. John W. Campbell's editorial "We 'Must' Study Psi" provides a fittingly controversial example. Campbell's essay, championing scientific exploration into mind control, telepathy, and divination, touched off a career-long project of advocating for what we might now consider "fringe science" in his editorials for *Astounding*. While Campbell's iron-handed approach to the content his writers produced made him controversial among SF authors, his willingness to consider pseudo-science and less-than-verified claims of psionic powers ultimately came to mark him as naive and easily manipulated in the eyes of many of his readers. Nevertheless, in this editorial, Campbell lays out the beliefs behind his interest in phenomena such as psychic forces and mind control, and documents his hope for human enhancement operating during the early part of the Golden Age. For Campbell, the superman stories he strongly encouraged his writers to produce, the editorials he authored on psionic research, and his participation in movements such as General Semantics and Dianetics all played important roles in the same project: advocating the scientific exploration of human potential.

The editorial opens with Campbell claiming that "the essential concept of truth-seeking is that a truth must be accepted, whether it is favorable or unfavorable, desired or dreaded, whether it means riches and happiness or stark madness."[46] Campbell is writing after having conducted his own studies of various psi-powered devices and reached the conclusion that there is a subjective force that can influence the objective universe in a profound way. He connects this force first to traditional beliefs in magic then to the concept of emotions. In both cases, Campbell suggests to his readers that, on these topics, "the very best advice Logicians, Philosophers, and Scientists have had has been . . . 'There shouldn't *be* any such thing! Suppress them! Deny them! Do away with them!'"[47] Surveying human history, Campbell finds that civilizations that have attempted to do away with magic and emotion have not been successful in their endeavors. He concludes from this that the universe can be divided into Subjective Reality and Objective Reality:

> I suggest that Subjective Reality bears the same relationship to Object reality that field-forces do to matter. Field forces are not material; they obey wildly different laws—but they do obey laws.
>
> I suggest that Subject Reality is a true, inherent level of reality in the Universe. It's no more something exclusively generated by human minds than 'organic' chemical compounds were exclusively generated by living organisms.[48]

Campbell suggests that psychic phenomenon are "the only objectively observable set of phenomena stemming from subjective forces."[49] In other words, Campbell wants to explore the subjective dimension of reality in order to improve our control over these psychic forces, as Subjective Reality is potentially a new, untapped realm of human ability. Moreover, the concept of "control" is important to Campbell's vision of expanded human potential, a word he repeats with great frequency in discussing these psionic forces:

> You can't control a phenomenon by denying its existence. You can't control it by suppressing it either; suppression simply causes an energy-storage effect that leads to eventual explosive release. If there's a river flowing through a

valley where you want to build a city, it's rather futile to simply build a dam to block the river; eventually the dam will be burst by the building pressure, and the city wiped out in the resultant flood.

A phenomenon can be controlled only by acknowledging it, studying it, understanding it, and directing it usefully. Properly handled, that river should be dammed, channeled through turbines, and made to supply the city with light and power.[50]

In addition to very potent imagery, the insistence on controlling psi, to dam the flow of energy to supply the world with "light and power," suggests a connection between Golden Age science fiction and cybernetics that is neither casual nor coincidental.[51] An insistent language of control suggests Campbell was interested in applying scientific principles to untapped, newly discovered human potential, thereby creating a kind of transhuman engineering. The focus on an applied science of human potential goes a long way toward explaining the interest many Golden Age SF writers, including van Vogt and Campbell, had in systems that promised to expand human potential.

Van Vogt's autobiography further intensifies these claims. Throughout he speaks of writing in a "non-human state" resembling "sleep," or more precisely, "a conscious not knowing what's next, dreaming about it, and then putting my brain to the conscious task of fitting it together."[52] His speculations on Cold War–era technologies take on a specifically evolutionary futurist tint as they echo Campbell's editorial:

I have a feeling that we're living in a world where no communication is yet satisfactory to anybody, because we all suspect danger. What's being said isn't good enough. It's an intermediate stage of history, and we're playing around with powerful toys. . . .

Human beings have basic abilities. These have nothing to do with the Morse Code or learning how to wage a war or writing a book. He's got something more basic than that, and we don't even know what that is. We're testing the wrong things. He has learning ability, and learning the Morse Code at such a speed merely proves that in this one area he has no barriers to using those basic abilities.

To describe human abilities, we use terms like creativity, intelligence, and so forth. These are not operational terms.[53]

Van Vogt argues in this section that there is nothing preventing humans five hundred years ago from learning to use contemporary technologies (outside of their not yet being invented). At the core, this signals that there is an adaptable potential in the human brain and all our scientific studies of humanity have failed to determine what this potential is. For van Vogt, "we have to notice that what we are doing is not in our control."[54] Writing later in his life, van Vogt reflects on the fact that the science of psi he and Campbell sought has not come to pass. Faced with nuclear destruction, we have to, according to van Vogt, develop a better means of asserting control over the forces of our desire.

Campbell and van Vogt diagnose this failure as a linguistic one, for they lament a fundamental failure on the part of vocabulary to describe human abilities accurately and creatively. Both men sought to parse out a better operational language for creating an engineering of human potential. Where earlier science fictions sought solutions to economic depression and massive unemployment in the form of technoscientific progress, the fictional supermen narratives and advocacy of real-world paranormal research documented throughout Campbell and van Vogt's careers represent, at the dawn of the Golden Age, the next stage in a politically progressive science fiction. In both the failure for progressive technocracy of GM's "Futurama" and the descent into a eerily Gernsbackian techno-fascism in Germany, humanity's ability to *comprehend* new technology rather than new technology itself became the chief problematic for science fiction's political project. Humanity's lust for power, destruction, and hierarchical thinking becomes, in the moment of critical technocracy's collapse, the new problematic science fiction chooses to work through. Campbell and van Vogt's engagement with psionic systems and their championing of superman fiction then become the means for building a technical discourse to address this problem, much like the critical technocrats sought a way to manage futuristic technology and the evolution of a progressive society. Consequently, while the psionic systems Campbell and van Vogt explore may seem naive and silly to us today, these fictional and real-world interventions were a response to the changing thinking in science fiction about the focus of advocacy and the best pathways toward a future available to humanity.

Boys with Their (Genetic) Toys

As an example of the influence of van Vogt and Campbell's experimentation with psionics, Philip K. Dick—who corresponded intensely with A. E. van Vogt while learning his craft as a young SF writer in the

1950s—suggests in "Exegesis," that the 2-3-74 experience "sort of makes me like a van Vogt character: possessing the most utterly priceless wisdom/formula-for-immortality on the planet," but he backtracks, fearing that "this is megalomania, for sure."[55] This misgiving reflects Dick's deep understanding of the technical construction of van Vogt's fiction, an understanding Brian Attebery uses in the only extant criticism on the SF Grandmaster's writing.[56] Attebery dismisses A. E. van Vogt as the embarrassing uncle to Philip K. Dick's more mature science fiction, but he also highlights the fact that engaging with van Vogt's extensive, almost singular collection of superman narratives was a crucial pathway into writing science fiction for Dick. Following the observation that Dick feels himself to be a singularly gifted human, destined to save the planet (the sort of naive wish-fulfillment narrative that so enraged Thomas M. Disch), he turns to discussing Richard Bucke's evolutionary futurist account of cosmic consciousness. From this, he concludes,

> In the past certain precursors of the New Man appeared (e.g., Socrates, Jesus). Dr. Bucke thought the frequency would increase soon. This ties in with Bergson's élan vital, too, and with Eros as the push of life forward in evolution. This is how God works. This is how God has always worked, from the day creation began: progressively, successively, continuously. "Day" after "day." Dr. Bucke's wise theory would account for the rarity of cosmic consciousness in the past, and would untie the knot of the dichotomy expressed above. I am, ahem, like a van Vogt character after all; like a Slan. (The next step up).[57]

While Dick may not be the brave and magical golden child of van Vogt's superman fiction, he does find succor in the idea that the vexing experience that tormented the final eight years of his life may result from the kind of evolutionary futurism outlined in van Vogt's myriad superman narratives, most famously and successfully in *Slan*. Moreover, this identification was not limited merely to Dick. Many SF fans during the 1940s felt strongly that they, like Dick, might be "like a Slan."

In this section, I will use *Slan* as *the* archetypal superman narrative, in order to capture some characteristics of the narrative type. Following Dick's example of identifying with these supermen, I highlight the fan response to *Slan,* which in addition to being immensely popular, also inspired a number of proposed and actual Utopian communities organized around the production and consumption of science fiction. These fan utopias, and the rhetoric they produced in support of their vision of fan *communitas,* will reveal the shifting nexus of political progressivism during this period of SF history. In addition to being interested in superman narratives like *Slan,* these fan utopians deployed evolutionary futurist rhetoric in the call for a fan-led evolutionary vanguard that would actualize the radical human potential being documented by Campbell and his writers in the pages of *Astounding.* Far from the isolated and awkward popular image of the SF fan promulgated in mass culture, these fan utopias document a fascinating engagement with the kind of inner transhumanism discussed in the previous chapter, while simultaneously merging with a techno-progressive vision that documents a more complete and thorough utopian program than presented in previous periods of SF history.

The 1939 novel is quite clearly a transition between Gernsbackian critical technocracy and the evolutionary futurist vision of humanity overcoming itself that marked the Golden Age's more "serious" social science fiction. As much as the novel glorifies Jommy's genetically enhanced abilities, it is equally obsessed with cataloging the various high-tech weapons Jommy builds for himself (Jommy's dead father is, tellingly, the kind of citizen-inventor so important to the populist science of the Gernsback period). Specifically, Jommy's car—the "very ordinary-looking, very special battleship on wheels which he never allowed out of his sight"—is an object of fetishistic description in the latter half of the novel, which documents Jommy's search for other slans.[58] His "long, low-built machine" has "some weapons that human beings cannot match," a coating of "ten-point steel" that "can stop the most powerful explosive known to human beings."[59] When he finally meets another slan, Kathleen Layton, he tells her that she only has to "stay within a few hundred yards of my car to be absolutely safe."[60] As can be surmised from these samplings, there is a lot of talk of Jommy's superior arsenal and futuristic weaponry in *Slan.* In fact, equipment fetishism is a unifying feature of many of van Vogt's novels, especially the lengthy descriptions of guns in the Weapon Shop novels (*The Weapon Shops of Isher* [1951] and *The Weapon Makers* [1952]). While, as Brian Attebery argues, van Vogt's fiction lacks mature sensibilities about sexuality and intimate human relationships, van Vogt's descriptions of his characters' weapons and futuristic gear are positively erotic.

In any case, these descriptions align with Disch's denunciation of a Gernsback-era science fiction as nothing but a "menu of future wonders . . . that boys like best—invincible weapons and impressive means of transportation."[61] At the same time that he writes with a quasi-pornographic lust about futuristic gizmos, guns, and rocket ships, van Vogt also uses the same language to describe the mental and physical "gear" of a genetically enhanced superman's body. For instance, just a few pages after the loving description of Jommy's car, Jommy and Kathleen begin to explore their telepathic links:

> What a rich joy it was to be able to entwine your mind with another sympathetic brain so intimately that the two streams of thought seemed one, and question and answer and all discussions included instantly all the subtle overtones that the cold medium of words could never transmit.[62]

In the same way that van Vogt's writing about cars and guns contributes to the Gernsbackian project of selling readers on the techno-wonder of a specific model of futurity, his writing about mental and physical superpowers attempts to sell his readers on the process of evolutionary futurism, marketing the rational approach to psionics van Vogt and Campbell were exploring in their daily lives.

This advertisement's focus on an evolved and persecuted minority resonated with fans in interesting and productive ways. Brian Attebery's reading of *Slan* dismisses these engagements as merely playing out the dominant "wish-fulfillment" model often used to relegate certain SF writers to the scrap heap of history. He sees the popularity of *Slan* as a form of Althusserian "hailing" that managed

> not merely to invite identification with Jommy Cross but to construct a particular identity for the science-fiction fan as slanlike superhuman in disguise. Not only did he hail the readers, for whom the phrase "fans are slans" became a byword at conventions, but also John W. Campbell himself. Campbell believed that he shared more than a set of initials with Jommy Cross.[63]

In Brian Attebery's dismissal of van Vogt as a hack, the superman narrative, archetypically established in the pages of *Slan*, becomes a structuring narrative for early fandom: isolated because of their differentness, slans and fans both need to remain hidden among the larger body of the human race. It is interesting, in Attebery's reading of this fiction, that he immediately interprets the trope of the isolated superman as an anti-utopian figure, reproducing the problematic reading of Nietzsche amongst the transhumanists discussed in chapter 1:

> From a political perspective, the Nietzschean superman is the antithesis of utopia. In utopian fiction, progress is a communal progress—the entire

> society evolves together. The mechanisms for improvement are generally institutions: schools, governments, families, political groups, media and so on. The utopian view parallels recent revisions of Darwin that hypothesize a selective advantage to cooperation within communities. In superman stories, however, it is every man for himself: the individual evolves apart from, or even in opposition to, his society. Only after he has separated himself, in the version authorized by Campbell, can he reenter the community as its anointed leader.[64]

While there is certainly some truth to this idea, especially as Jommy wishes to use his telepathy to control the minds of a hateful world in order to gain acceptance, the larger evolutionary utopianism of *Slan*, and much of van Vogt's fiction, documents a persecuted minority as a kind of evolutionary vanguard, eventually pointing the way toward a more broadly shared enlightened way of being. Just as Gernsbackian wonder fiction raised the specter of an enlightened technocracy, these narratives suggest a genetic evolutionary destiny greater than present conditions; however, it is also important to reiterate that, as happens throughout *Slan*, organization of these disparate individuals is key to unlocking this evolutionary destiny. While this may smack of a rhetoric of fascistic master races, the superman narrative attempts to walk a careful line, as we shall see, between telling what to do and nurturing nascent evolutionary potential.

From Slan Shack to the Raven's Roost: Fan Utopias during World War II

As Attebery suggests, "fans are slans" was an important rallying cry of early fandom. Kevin J. Anderson's introduction to the most recent reprint of *Slan* expands on this claim:

> "Fans are slans" was actually a slogan used by fledgling SF fandom in the 1940s, a group of whom founded their own cooperative housing development in Battle Creek, Michigan. They dyed streaks–surrogate tendrils–in the hair at the back of the head and moved into an eight-room house that they called the "Slan Shack."[65]

This cooperative housing development and its formation in Battle Creek points to a much more interesting and politically rich history of fan utopian communities during the 1930s and 1940s than suggested by Attebery's antipolitical account of the isolated, sexually frustrated, and hopelessly nerdy stereotype of *Slan*'s readers. These real and imagined intentional communities—all dedicated to fandom that dominated fan discussion during the early days of the Golden Age—I call "fan utopias."[66] These existing intentional communities, and the grand plans for future fan settlements, suggest a much more profound engagement with the superman narrative presented in *Slan* than mere wish fulfillment.

The Slan Shack was the name of a real-life house[67] in Battle Creek, Michigan, inhabited starting in 1943 by SF fans Al and Abby Lu Ashley, Walt Liebscher, and Jack Wiedenbeck.[68] When not manufacturing explosives in a munitions factory in Indiana, E. E. "Doc" Smith—the author of the *Lensman* novels and the inventor of space opera—was also a resident, as he had worked nearby before World War II. More important, "Slan Shack" provided the collective name to a whole host of fan houses inhabited by both SF fans and SF authors during the 1930s and 1940s, as these dwellings became referred to as "slan shacks" in the history of fandom.[69] These communal houses were incubators for early fan culture but were also, notably, a response to the general housing crisis associated with the mass migration of workers and soldiers as the United States' extensive wartime economic build-up began operating.

The Slan Shack derived its name partly from having been the birthplace of the "fans are slans" slogan. According to Dal Coger's history of Slan Shack, at a party at the Ashleys' house in Battle Creek,

> We all reveled in fan talk and someone came up with the idea, "Hey, wouldn't it be great if we could get fans together and have our own apartment house?" A. E. Van Vogt's *Slan* had been published a year or so earlier as a serial in *Astounding* and someone had almost immediately asked, "Do you suppose fans are Slans?" (Meaning, were we a mutation from the mundane variety of humans? No one took the idea seriously, of course.) But our idea of closer association was promptly named Slan Center.
>
> Our planning included a fanzine room where all occupants would share access to a mimeo, and apartments with northern light for the artists (Jack W.'s idea). What was behind this was the feeling of closeness, of being able to be open in our ideas, that we as fans could express most easily in each other's company. Everyone had experienced the raised eyebrows of mundanes when you tried to discuss science fictional ideas with them. Slan Center would make it possible to be openly fannish any time we were away from work.[70]

These impulses for coming together were part of the generally changing nature of fandom, as chronicled, for example, by Jack Speer in his fan history *Up to Now* (1939). Speer's account of fandom during the 1930s reveals the origin of the numerical historical periodization commonly used in other fan histories. Speer argued for a history of fandom occurring in waves, breaking the history of fandom during the 1930s and 1940s into three numbered eras. Speer argues that first fandom was organized primarily through the letter columns of pulp magazines by Donald A. Wollheim, who wanted to turn fandom into an organized affiliate of the American Communist Party. Following Wollheim's expulsion from fandom (a theme that, as we shall see, repeats throughout early fandom), second fandom "was marked by a shift of interest away from the pro field (then in recession) to the fans themselves" through organizations such as Slan Shack and also the first science fiction conventions that emerged in the late 1930s.[71]

This newly emerging and coherent fandom was no longer content, as was first fandom, to communicate with one another in the letter columns of professional magazines or through amateur press associations such as Fantasy Amateur Press Association (FAPA); it was beginning to organize into a distributed affinity community who increasingly wanted to spend time together. Following the narrative patterns established by *Slan,* this fan community regarded itself as an enlightened minority within a broader human world. Harry Warner Jr., in his fan history *All Our Yesterdays,* quotes Al Ashley as saying at the time of Slan Shack's creation that "the average fan enjoys intellectual superiority over the average man. But that only means that as a select group we excel the human average. No effort would be needed to find other select groups which surpass the fen intellectually."[72] ("Fen" was an early plural of "fan" in the language of SF fandom.) Here, Warner seems to suggest that some early fans, in the example of Al Ashley, felt that science fiction fandom may have actually represented a new evolutionary leap in humanity. In fact, Warner continues by quoting an unidentified piece by Jack Speer: "Practically all fans fell into the upper one-quarter of the population in intelligence, and the average is within the top ten percent."[73] Later in life, in an interview, Harry Winter also observed:

> I think [*Slan*] may have inspired a lot of individuals at the time into thinking that maybe fans were a "chosen race" because the Slans in the story were separate and different from the rest of humanity, and fans in those days did feel a sense of being "different" somehow.[74]

This view fans took, of a chosen race of human evolutionary elites, partly informs the creation of the Slan Shack (and other subsequent fan utopias), although it is important to note that fans were also drawn to cohabitation due to a perceived outsider status. Harry Winter, discussing the Slan Shack phenomenon, observed:

> [Jack] Speer thinks that all fans are handicapped in some way or another but he has to stretch "handicapped" to cover so many different circumstances that I don't think his theory holds up. For instance, he thinks growing up in a small town is a handicap.[75]

So in addition to being more intelligent, fans, like Jommy in *Slan*, felt persecuted and isolated due to their intellect. As Dal Coger observed, "we all seem, in retrospect, to have been a bunch of misfits looking for a niche in society."[76]

Following this need and the formation of Slan Shack and other, similar fan communes, the Ashleys and their cohabitants began to take seriously a larger project: Slan Center. This idea was to have been a much broader, intentional community to be formed outside of a city, originally Battle Creek but later Los Angeles, once the Ashleys and Slan Shack moved West after World War II. It was to be composed of a number of houses and apartment towers to be structured around a shared printing facility (for the production of fanzines). As Warner points out:

> Fans can get along well with one another in such instances as Berkeley, and there is no intrinsic reason why fans should not make up the population of a city block, if they can run a household. Ashley suggested a location on the outskirts of a large city which would contain "a collection of adjacent individual dwellings sprinkled with a few apartment structures and with a large communal building." Choice of the site would be made with an eye to the city's current fan population, to permit some of the city's current fan population, to permit some of the centre's inhabitants to avoid a drastic break with familiar surroundings. And it should be understood that this proposal was taken very seriously by level-headed fans, at the time it was made.[77]

The most interesting point to note about the above selection is Warner's assertion that this proposal was taken seriously by "level-headed fans" of the era. The first proposal for Slan Center, as well as the first account of Slan Shack, appeared in Al and Abby Lu Ashley's fanzine, *En Garde*, in the June 1943 issue. A striking feature of the Ashleys' rhetoric in this piece is the tentative nature of their word choice. The idea of fan cohabitation, in an intentional and organized manner, was "rather timidly advanced" by the Ashleys to fans Niel DeJack and Dal Coger during the visit described above, an idea the Ashleys "had been toying with in our own mind for some time."[78] They also mention that when the idea was first discussed at Midgicon in Chicago during spring 1943, "we all had an excruciating lot of fun" twisting it "into the most humorous brainstorm of the year . . . yet when the laughter finally subsided, the sincere desire to try the project remained."[79] The result of this "sincere desire" is the Ashleys' piece in *En Garde.*

In making the case for Slan Center, they argue for the creation of a housing cooperative in Battle Creek (and later, as mentioned above, discussed in terms of a Slan Center to Los Angeles). They suggest that

> A suitable square block could probably be purchased for from three to five thousand dollars. Then an ultra modern group of homes, apartments, housing units, or whatever you wish to call them, would be built around the block. In the center, formed by the square (in the collective backyard, if you will), a larger communal building would be constructed. This would serve as a meeting hall, library, publishing headquarters, central heating plant, and even an electric plant. If desirable, there could even be a small machine and woodworking shop for those who enjoy such hobbies. While each unit of the project would be distinct, the complete group of structures would be designed to form a pleasing architectural whole.[80]

The particularly interesting aspect of this very prudent plan, especially when considering the idea of taking science fiction seriously, is its inclusion of central heating and an electric plant. The Ashleys imagine Slan Center to be a self-sufficient community and, more important, are even imagining mundane details of this community; their piece in *En Garde* goes beyond speculation or wish fulfillment and focuses on practical solutions for creating a fan community. Beyond self-sufficiency, the Slan Center would focus on intellectual growth. Despite some of the "silliness" associated with fandom, the Ashleys suggest that in fan interaction, still rare during these early days, "there is also a very satisfying exchange of ideas which, in the end, overshadows all else."[81] Additionally, the Ashleys, in making the case that fans are indeed slans, argue that "the real fan is mentally wide-awake. He is readily able to accept new concepts, and his mental tastes far exceed those found in the ordinary person of comparable age."[82] They close their argument for Slan Center by suggesting that "in union there is undoubted strength. The day may come when fans are a group to be reckoned with."[83]

This argument for the possibility of collective overcoming and advanced futurist thought was, according to numerous fan histories, considered for a number of years, and more than a few fans were considering the idea of an organized Slan Center as crucial to the organization of postwar fandom. So, why is America not awash in fan collectives?

The answer to that question is "Claude Degler."

Damned Degler! Fascism and the Failure of Fan Utopia

Claude Degler was an itinerant active in fandom from around 1943 to 1947. Traveling around the country as a hitchhiker, Degler served as the Johnny Appleseed of early SF fandom. His entry in *Fancyclopedia 2* lists his involvement in organizing fan groups throughout the southeastern United States, in New York City, Philadelphia, Hagerstown, and Oklahoma. All of this organizational work was in service of the grander vision of SF fandom that Degler called "The Cosmic Circle." Beyond his wide-ranging travels, he also published a torrent of mimeographed fanzines to proselytize for his vision of a union of cosmic fandom. Degler's organizing activities and his larger vision made him an important, though controversial, figure during World War II–era fandom.

Similar to Ashley's interest in using Slan Center to organize fans, Degler's Cosmic Circle sought "to contact as many persons in all countries who have interests similar to ours" and create a global, recognizable fan organization.[84] Degler, who had a 4-F draft status prohibiting him from service, specifically articulated that it was the duty of fans on the home front to organize because "fans returning from the war will want to find a well organized *United Fandom* they can be proud to rejoin."[85] Moreover, Degler recognizes, in his call for a unified fandom, fandom as not just an activity but a way of life, calling for fan-oriented options for all sorts of cultural activities, including "literature, music, history for *united fandom* . . . books for fan children."[86]

He refers to this comprehensive vision as "fanationalism" in the first issue of *Cosmic Circle Commentator.* This unified and organized global fan culture held the possibility of transitioning fan interest in fictional issues such as space travel into a force that could "exert our influence in building that kind of a world we most want," as he articulated in the first issue of *Cosmic Circle Monthly.*[87] Degler connects this new and better world to evolutionary futurist rhetoric at numerous points in his work. In addition to fashioning and presenting a strong and unified fandom to market science fiction as a lifestyle to the general public, he envisioned the Cosmic

Circle specifically as a means of "promoting cosmic consciousness," by which fandom would be transfigured as a means to a genetically superior humanity. His utopian project for fandom also mirrors the call for a Slan Center–style development, with a suggestion of a Cosmic City for fandom. Additionally, Degler mentions the apparently real "Cosmic Camp . . . a tract of land in northwestern Arkansas. This tract in the mountains of Van Buren County, near fishing, hunting, and other recreational facilities, may be used free by vacationing members of *Cosmic Circle*."[88] This Cosmic Camp would become infamous in fan circles during the period marked by Degler's heavy campaigning.

Regarding Cosmic Camp, Warner deadpanned that it was a "wilderness settlement in the Ozarks where fans would make love and rise above humanity."[89] What separated Degler from the Ashleys was his interest in the active production of genetically superior beings in the real world. The reproductive agenda of the Cosmic Circle, more than anything else (other than the fact that many fans who knew him described him as "annoying") was what doomed Degler's plans for fandom. While cohabitation was perfectly acceptable to fans of the era, Degler's connection between mind expansion and free love came at least twenty years too soon. The sexual content of the Cosmic Circle opened up Degler's utopian vision to attack from fans, who, while perfectly willing to believe themselves to be genetic supermen, found Degler's methods for actualizing this potential too esoteric (or otherwise challenging) for people who spend their time reading about space travel and robot overlords.

This difference between Degler and mainstream fandom can be seen in the later parts of issue 1 of *Cosmic Circle Commentator,* which collects excerpts from Degler's more philosophical *Cosmic Circle Digest.*[90] These excerpts are very different from the more levelheaded tone of the rest of *Commentator* and explain more of both Degler's vision for fandom and Warner's deadpan about making love and rising above humanity. Degler offers a "Declaration of Existence: of a new race or group of cosmic thinking people, a new way of life, a cosmology of all things. Homo Cosmens, the cosmic man, will appear. We believe that we are actual mutations of the species."[91] Following the narrative pattern of superman fiction, Degler writes that "we are convinced there are a considerable number of people like ourselves on this planet, if only we locate them. Someday we'll find most of them, and then we'll do great things together."[92] In juxtaposing these two quotes, we find a more explicit connection between the Nietzschean superman and the supermen of van Vogt's narrative pattern. Based on Harry Warner Jr. and Dal Coger's fan histories of the Degler period, Degler was interested in using the Cosmic Circle as a literal and figurative breeding ground for future supermen. Degler's more cosmic and more explicitly transhuman vision of fandom did not sit well with many fans, despite the numerous members in his Cosmic Circle during the 1940s. Degler's major mistake seems to have been too take too seriously the declaration that "fans are slans," identifying *Slan* not as a metaphorical evocation of outsider status but *literally* about the genetic status of fandom and a coming cosmic posthumanity.

It is difficult to sort out a lot of Degler's history for two reasons. In addition to being a cosmic and transhuman visionary (as well as, it is possible to suggest, a proto-hippie), Degler was also partly a P. T. Barnum–esque huckster. His constantly shifting journals, published under various pseudonyms, sought to inflate his ethos and manufacture himself as the representative of a vast fan empire that, for the most part, did not exist. Additionally, in a manner not unlike Stalinist historical revisions during the same period, fandom wrote Degler out of its history. Fans broke with Degler, rather violently, insinuating in later publications such as Jack Speer's *The Cosmic Circle and Fandom* (which was distributed to editors of professional SF magazines and members of the news media who may have received Degler's journals, and explained that the Cosmic Circle was not representative of fandom) that Degler had had sexual relations with underage women in his hometown of Newcastle, Indiana, and had spent time in a state mental facility. The *Fancyclopedia 2,* itself a one-sided publication in which the victors wrote the history, argues that fans tired of dealing with Degler, who was mentioned on numerous occasions in fan histories as having shown up uninvited at peoples' houses and was often described as "mentally unstable." As a result, fans began to publish parody fanzines such as *Trivial Triangle Troubador* and *Comic Circle Commentator.* Eventually, Degler's memberships in FAPA and the Los Angeles Science Fiction Society were revoked—tantamount, during this period, to being kicked out of fandom.

The *Fancyclopedia 2* entry does proffer an interesting explanation for fandom's sudden rejection of, then subsequent post-facto hatred toward, the Cosmic Circle:

> [A] copy of the *Cosmic Circle Commentator* had come into the hands of *Amazing Stories*' editor [Raymond A.] Palmer. The declaration of existence of a super race smelled to him of Naziism, and the fanationalistic program

seemed the horrid ultima of fans' movement away from the pros which he, as a fan of the First Fandom and now as a frankly commercial editor, decried. Because of this, and because fans were not the type of readers his publications catered to, he made it known thru *FFF Newsweekly* that fans or fandom would not get into the letter departments in the future, originals would not be contributed for auction at fan gatherings, and so on. Some fans reacted by saying that Degler's ideas in some form had all been spoken in fandom before, and who the hell was Palmer to try to dictate to fandom, and as for *Amazing Stories* and *Fantastic Adventures*, good riddance of bad rubbish. Others, alarmed at the possibility that other pros mite [*sic*] follow Ziff-Davis's lead and cut fandom off from financial, recruiting, and publicity assistance, made haste to inform Palmer that Degler didn't speak for fandom. Palmer modified his statement of the ban, but urged fen to return to the ways of their fathers.[93]

The connection to Nazism regarding Degler's journal is particularly interesting. It was around the time of this incident (1943) that professional interest in supermen narratives began to cool down. The revision of Degler into a minor figure in the history of fandom may have resulted from the threat he ultimately posed to fandom's survival. In addition to the numerous mentions (in Degler's and other fanzines from the period) of the depletion of fandom due to the draft, Degler's increasing association with the very organization America was at war with made him increasingly untouchable in fan circles. While the superman narratives van Vogt helped to pioneer argued for an enlightened and benign evolutionary vanguard, Degler's plans to actualize that genetic future through a program of selective fan breeding highlighted the latent rhetorical connections to Nazism at the core of the entire superman boom. While themes of genetic supermen persisted in science fiction (and to some extent still persist today), the rhetorical strategies in which these supermen were marketed to SF readers changed. This mutation in the narrative code may have been partly due to this incident between Palmer and Degler.

In any case, Palmer's threat to blackball fans from his letter column, by association, also dissolved plans, however real they may have been, for Slan Center. Degler, in organizing for a cosmic consciousness and attempting to create actual, existing supermen, took science fiction *too* seriously, in contrast to the Ashleys' more reasonable appropriation of the themes of *Slan*, or Campbell and van Vogt's own levelheaded psionic engineering. The Cosmic Circle incident also served to stiff-arm the emergence of fandom as a distinct and utopian cultural force outside the fiction around which it had organized. Palmer's threatened boycott of fans would have, as the above selection suggests, removed key financial support for a moment in fandom "marked by a shift of interest away from the pro field (then in recession) to the fans themselves."[94] The reminder of fandom's dependence on letter columns and classified ads in the professional pulps served to check a fandom that was, during this period, relatively autonomous. Fandom during this period almost emerged as a distinct and distinctly utopian manner of evolutionary futurist being, with a fan identity that extended beyond merely an interest in reading science fiction.

These fan utopias—Slan Center and the Cosmic Circle—document an important moment of reality–fiction interchange in the history of science fiction, when fans were beginning to move beyond the consumption of science fiction and into the creation of a unique and autonomous futurist culture. Given Degler's ejection, and given that it was his fanzine that provoked Palmer in the first place, he is widely credited with killing off the idea of Slan Center, as well as his own more boffo plans. While Degler is the one who carries the blame, he was not the only one making crypto-fascist utterances in calls for fan cohabitation. In the Ashleys' call for a Slan Center, they suggest that "in union," such as that offered by fandom, "there is undoubted strength. The day may come when fans are a group to be reckoned with."[95] Written two years into America's involvement in the struggle against global fascism, the echoes of Nazi sloganeering in this assertion is as alarming as anything Degler wrote. Palmer's intervention into vaguely fascistic and eugenic discourse in SF fandom during the 1940s was a major corrective to the creation of an autonomous fan futurism, but Palmer also importantly reminded fandom of the serious stakes of any speculative operation.

Seriousness as a Problem of the Future

Having seen how Claude Degler's enthusiasm for a proto-transhuman fan culture signaled the collapse of a utopian and autonomous SF fan culture during World War II, the questions animating this chapter remain. What does it mean to take science fiction seriously? And why is it a problem? I want to consider Degler's ejection from fandom as a test case for the limits of a rhetoric of seriousness within evolutionary futurist discourse. As we have seen, Degler's plan for the Cosmic Circle violates some unspoken rhetorical decorum, an unspoken injunction against a specific kind of seriousness. As we have seen, Degler's ideas are only slightly afield from those of the organizers of Slan Shack, yet his ideas appeared dangerously radical, while Slan Shack was consistently remarked on as a "level-headed" scheme.

Part of this story lies in the distinction Richard A. Lanham makes between rhetoric and seriousness. In his exploration of Renaissance rhetoric, *The Motives of Eloquence* (1976), Lanham offers a helpful distinction in analyzing the seemingly endless turf war between rhetoric and philosophy throughout the history of Western ideas. This dispute hinges on the nature of linguistic style and, as Lanham characterizes, it can be thought of as a fight between two incommensurate worldviews that bear with them two fundamentally incompatible versions of what humans are and what humanity does. Lanham suggests that style, in the philosophical worldview, is all about seriousness,

> The discussion of verbal style in the West has proceeded on the basis of a few simple premises which it may help to hold before us. I shall call them serious premises. They run something like this. Every man possesses a central self, an irreducible identity. These selves combine into a single, homogeneously real society which constitutes a referent reality for the men living in it. This referent society is in turn contained in a physical nature itself referential, standing "out there," independent of man. Man has invented language to communicate with his fellow man. He communicates facts and concepts about both nature and society. He can also communicate a third category of response, emotions. When he is communicating facts or concepts, the success with which they are communicated is measured by something we call clarity. When he is communicating feelings, success is measured by something we call sincerity, faithfulness to the self who is doing the feeling.[96]

In contrast to Lanham's very serious work of fidelity to the self we find, of course, rhetoric. Arguing that "the Western self has from the beginning been composed of a shifting and perpetually uneasy combination of *homo rhetoricus* and *homo seriosus,* of a social self and a central self," Lanham suggests that *homo rhetoricus* is committed to undercutting the foundational philosophical assumptions of *homo seriosus.*[97] For Lanham, this distinction is the basic terms on which rhetoric and philosophy will fight with one another for the next few thousand years.

Lanham leverages Plato's depiction, in several of his dialogues, of the split between Socrates, the preeminent example of *homo seriosus,* and the Sophists as a template for the manner in which this distinction has fugued throughout Western history. Suggesting that *homo rhetoricus* is not committed to any position beyond winning, and that, furthermore, it possesses no "single value-structure," Lanham further characterizes seriousness as a stance meant as a bulwark against the reality implied by the partiality of the rhetorical man's position.[98] For Lanham, humanity desires this bulwark:

> We would prefer to dwell on our tragic fate, painful but heroic. To set ourselves off against the whole universe makes us, in a manner of speaking, as big as it is. *Homo rhetoricus* is flung into a meaningless universe too, of course. But unlike his serious—or existential—*doppelgänger,* he doesn't repine, bathe in self-pity because his world possesses no center. He can resist such centermentalism because he knows that his own capacity to make up comforting illusions is as infinite as the universe he is flung into. Naked into the world he may come, but not without resource.[99]

Lanham suggests that rhetoric is committed to partiality and to pragmatic language usage because, unlike the serious ethos of the philosopher, rhetoric presents a worldview in which we do not care that our lives have no meaning. In this model, meaning is beside the point. Seriousness is a trap that stands in the way of getting things done.

The most provocative aspect of Lanham's discussion is the rapidity with which his argument moves from a discussion of a dispute over the use of rhetorical style (serious man doesn't have it and doesn't need it; rhetorical man is all style and no substance) to the very nature of reality itself. For Lanham, the rhetorical viewpoint holds that "reality is what is accepted as reality, what is useful."[100] Moreover, Lanham argues that "rhetorical man is trained not to discover reality but to manipulate it."[101] That Lanham hinges his distinction between seriousness and style on the question of discovering reality versus manipulating it shows how much the "future" of evolutionary futurism manifests as a rhetorical problem.[102] While Claude Degler (at least the Claude Degler who survives in the negative portrayals in fan histories) exudes something of the sophist in his ability to manipulate the open networks of fandom to promulgate his ideas regarding cosmic futurism, his sustained commitment to an evolutionary futurist vision of human overcoming cuts through much of the faddishness of the period in fandom and suggests a true conviction to his cause. Degler, like the other transhumanists we have been discussing, marshals language to manifest a shared vision of a new future humanity. In other words, he is neither manipulating a shifting reality nor defending a transhistorical one. He creates a new reality through a rhetorical vision of the future.

In order to account for this creation of a future, a new set of distinctions beyond seriousness and style are needed. While Degler certainly plays around with both registers, in terms of thinking about Utopian constructions of the future he has one perspective. Lee Edelman's account of queer futurity, *No Future* (2004), offers an alternative set of terms for thinking through the manner in which both modes, seriousness and its opposite, are engaged in projecting futures rhetorically. Edelman identifies an organizing principle in the social that he calls "reproductive futurism," which works by invoking "terms that impose an ideological limit on political discourse as such, preserving in the process the absolute privilege of heteronormativity by rendering unthinkable, by casting outside the political domain, the possibility of a queer resistance to this organizing principle of communal relations."[103] He goes on to clarify that,

> for politics, however radical the means by which specific constituencies attempt to produce a more desirable social order, remains, at its core, conservative insofar as it works to *affirm* a structure, to *authenticate* social order, which it then intends to transmit to the future in the form of its inner Child. That Child remains the perpetual horizon of every acknowledged politics, the fantasmatic beneficiary of every political intervention.[104]

Edelman analyzes various contemporary forms of reproductive futurism, showing the pervasiveness of "for the children" arguments that attempt to reproduce a future of absolute sameness through heterosexual reproduction and childrearing. In a discussion that compares Walter Benjamin and the Third Reich, Edelman claims that it "is not to say that the difference of those political programs makes no difference, but rather that both, *as political programs,* are programmed to reify difference and thus to secure, in the form of the future, the order of the same."[105] Reproductive futurism then becomes a rhetorical model for the future that continually reinscribes present understandings of "normal" through the figure of the child. Edelman, in contrast, situates queer sexuality, which is nonreproductive sexuality, as a counterproposal to reproductive futurity. To borrow a chapter title from Fredric Jameson's *Archaeologies of the Future,* queer futurism is the "future as disruption," a rhetoric of the future that privileges the inscription of something else, something potentially new.[106]

Edelman's distinction between a serious future and a queer future highlights that both of Lanham's positions project futures:[107] *homo seriosus*'s desire to preserve a stable and knowable world into the future and the more polyglot positions of *homo rhetoricus* within a world of shifting norms, values, and programs, the unstable world of language itself. Thus, we can argue that the future is primarily a rhetorical construction. However, is it the product of style or seriousness? This question is at the core of Fukuyama's statement about transhumanism and science fiction. Putting our faith in stylistic products seems, according to Fukuyama, to be the wrong kind of seriousness. Putting our faith, however, in science and engineering (not to mention bioconservative morality) is a seriousness we should endorse. To take science fiction seriously, according to Fukuyama, is to do seriousness wrong. In one final dispute, between the Apollonian and the Dionysian in Nietzsche, we can fully understand the stakes surrounding this complicated rhetoric of seriousness that Fukuyama evokes.

In advance of *The Birth of Tragedy* (1872), Nietzsche wrote "The Dionysiac Worldview," making it the first text, as philosopher Claudia Crawford explains, "in which he clearly distinguishes the aesthetic categories of the Apollonian and Dionysian which underlie" the subsequent text.[108] This distinction, between "two deities" that "established . . . the double source of [Greek] art," also served as a watershed in Nietzsche's maturation as a philosopher. In the present context, they also serve to shape our understanding of Fukuyama's claims about seriousness and the transhuman. On the one hand, Fukuyama dismisses transhumanism for being nothing but science fiction taken too seriously, but on the other, he and his fellow bioconservative policy makers take transhumanism as a deadly serious threat to the very future of humanity itself. This chapter has addressed the first of Fukuyama's claims of seriousness, showing that science fiction has always been about taking itself very seriously as a means for imagining and addressing new, alien forms of technosocial order. I hope that focusing on the Dionysian content of evolutionary futurist rhetoric, embodied in the figure of Claude Degler, will tackle the matter of the second.

In Nietzsche's distinction, Apollo and Dionysos represent two modes through which Greek art creates experience. These two modes map to two exceptional psychic states produced through art: dreaming and intoxication. Apollo, for Nietzsche, is a god of art "only inasmuch as he is the god of dream-representations," governing the translation of "the higher truth, the perfection of these dream-states" into representations within "the only partially intelligible reality of the daylight world."[109] However, this God, "who reveals himself in brilliance" and for whom "'beauty' is his element, eternal youth his companion," demands certain strict behaviors in his artistic adherents.[110] Nietzsche stresses that this god of beauty, brilliance, and the representative image

> must also include that delicate line which the dream image must not overstep if its effect is not to become pathological, in which case the semblance does not simply deceive but also cheats; it must include that measured limitation, that freedom from wilder impulses, that wise calm of the image-making god. His eye must be 'sun-like' and calm; even when it is angry and shows displeasure, the consecrated aura of lovely semblance surrounds it.[111]

There is an element of decorum in the ordered beauty of Apollonian art that must not be transgressed. The artist can take his claims only so far, cut off from "wilder impulses" and governed by "measured limitation." The Apollonian structures the Ashleys' proposal for Slan Center, presented as it is with words like "timid" and "level-headed." Slan Center was a modest proposal that fans might band together through a shared interest in science fiction and weather the Great Depression together, in all seriousness. And if some genetic supermen might result from interbreeding among the supposed superior intellects of fans, all the better. What's a few next-stages-of-human-evolution among friends?

As we saw, Degler threw this modest scheme into chaos by, like Francis Fukuyama critiquing transhumanism, simultaneously taking "fans are slans" too seriously, yet not seriously enough. Degler's cosmic vision, amplifying the implicit supermen argument present in *Slan* and its more "level-headed" adherents, highlights the dangerous implications for taking seriously the experimental overcoming of the very category of the human: namely, the idea that to make slans is to breed. This eruption of nonnormative sexuality, focused as it is on producing a new future and not a reinscription of the same, provokes the distaste with which Degler's proposals were met. This horror, I think, is the true antithesis to seriousness here, not necessarily the triviality Fukuyama wants to imply by suggesting that transhumanism is science fiction taken too seriously.

Moreover, this horror can be seen as an effect of a Dionysian rhetoric shaping the foundation of ideas of human overcoming. In contrast to the Apollonian, Nietzsche offers the Dionysian: a mode of artistic experience related to the exceptional psychic state of intoxication. For Nietzsche, Dionysian art "is based on play with intoxication, with the state of ecstasy."[112] As we saw in chapter 1, ecstasy is a watchword for a kind of cosmic consciousness, and this is certainly also the case here. Primarily experienced in "the drive of spring and narcotic drink," the Dionysian is the aesthetic of the revel, the deliriousness of losing oneself in the crowd. Following the kind of magical evolution we have already seen, Nietzsche suggests that, in the Dionysian, "the *principium individuationis* is disrupted, subjectivity disappears entirely before the erupting force of the general element in human life, indeed of the general element in nature. Not only do the festivals of Dionysos forge a bond between human beings, they also reconcile human beings and nature."[113] In this image of an ecstatic revel, we revisit a rhetorical mode we first encountered in Claude Degler. His intoxicated, feverish rhetorical construction of a cosmic humanity, together with his coupling it with the more "level-headed" rhetoric of the Slan Center, highlights his own commitment to the Dionysian: the energy animating his exuberant prose in his various publications.

Further, Nietzsche specifies that in the ecstatic reconciliation of "human beings and nature" characteristic of the Dionysian mode, "now the essence of nature is to be expressed; a new world of symbols is needed; the accompanying representations acquire a symbol in the images of an intensified human being."[114] We have already seen in P. D. Ouspensky and Mina Loy the intertwined nature of failure of the human and the failure of language, but Nietzsche intensifies even this claim: the loss of speech in intoxication intensifies the image of the human, signaling the possible becoming of something beyond the human. As Nietzsche suggests, in Dionysian art, "a nobler clay, a more precious marble is kneaded and chiseled here: the human being."[115] Through its experimentations in the shaping of a future humanity, we can align transhumanism with Dionysos, not Apollo.

Following Nietzsche, Lanham, and Edelman, the Apollonian is the serious. It is the realm of reproductive futurism, dedicated to the perfection of things as they are. This Apollonian rhetoric is also that of the bioconservative, that great enemy of the transhuman. This suggests why Fukuyama's dismissal of transhumanism's seriousness, while also proselytizing for taking it seriously, is actually correct. No matter the degree to which evolutionary futurism comes dressed in Apollonian discourses of bioethics, human rights, and risk management, the proposition of overturning the human is inherently Dionysian: the future of humanity as disruption rather than sameness. This disruption is why science fiction's "court jester" is so important to understanding transhumanism. Contemporary transhumanism is itself a kind of science-fictional realism, a combination of philosophy and fantasy in an emerging discourse of a new humanity. Fukuyama's double assertion then signals that our society is undergoing a staggeringly transhuman tropic realignment. The uncanny realities projected by science fiction are becoming not only the *topoi* of our democracy but, indeed, the basic building blocks of our reality. We can conclude that Fukuyama is right, though for the wrong reasons: transhumanism *is* science fiction and science fiction is to be taken very seriously indeed.

3 TOWARD OMEGA

Hedonism, Suffering, and the Evolutionary Vanguard

In an essay framing her larger cosmopolitical project,[1] Isabelle Stengers uses the figure of the idiot as a means to imagine a utopian project for the present. She suggests that the idiot

> is the one who always slows the others down, who resists the consensual way in which the situation is presented and in which emergencies mobilize thought or action. This is not because the presentation would be false or because emergencies are believed to be lies, but because "there is something more important." Don't ask him why; the idiot will neither reply nor discuss the issue.[2]

For Stengers, the idiot, of whom Herman Melville's Bartleby (who would prefer not to) is a potent example, is a figure beyond opposition: it is not that idiots disagree with particular programs; they would just prefer to be left alone. It is these withdrawing figures whom Stengers suggest those interested in Utopianism should most emulate. In the face of those who want to create a "good common world," she suggests that "the idea is precisely to slow down the construction of this common world, to create a space for hesitation regarding what it means to say 'good.'"[3] Stengers's idiot, then, is the one who perpetually doubts without content, "those who think in this climate of emergency, without denying it in any way but nonetheless murmuring that there is perhaps something more important."[4]

This "something more important," Stengers's recurring mantra, is the means to articulate an oppositional discourse in the present. One theme of contemporary transhumanism drawn out in the Introduction to this volume is that of speed: humanity moving faster into a slick, machinic embodiment. As Stengers maps with the idiot and the cosmopolitical, the world is currently oriented around and rebuilt by a conceptual figure opposite in every way from the idiot: the entrepreneur. The entrepreneur encompasses more than simply the valorized heroes of a Silicon Valley start-up ecology for Stengers. Instead, entrepreneurship is a way of being human that narrows the wide range of human activities to an extremely limited set of operations. Particularly, for Stengers's entrepreneur, there is no human growth outside the creation of new things "and emergence is nothing other than the consequences of the factual obstacles that they constitute for one another."[5] As she suggests, these entrepreneurs "may be open to whatever makes them advance, but only in so far as it makes them advance. They are persons of 'opportunity,' deaf and blind to the question of the world that their efforts contribute towards constructing."[6] This world of entrepreneurs—in which "like Über but for food" or "Twitter but for your photocopier" count as innovative business models—is one shaped by radical self-interest and one without self-consideration. This lack of consideration opens up an already existing world dominated by the transhuman figure of speed, in which doing things and looking busy is the only mode of recognized activity.

This focus on speed and lack of critical engagement is why Stengers injects her figure of the idiot: the idiot always simply says "no" and this "no" (without reason, without qualification, without enthymeme) is a resistive act within cybernetic cultures that works against what Paul Virilio has called the "green light of implicit consent."[7] Virilio bases this idea of saying nothing as saying "yes" on the system used by NATO to coordinate the bombing of Yugoslavia in 1999. Targets were proposed, and if there were no objections from the member nations, the target was bombed. As Virilio points out, this means that silence is assent, which runs counter to the ways in which silence has historically been read within dialogue. Thus, for Stengers, working within this same space that views speed as a virtue and critique as weakness, the idiot becomes a figure of political (dis)engagement: by simply saying "no" the idiot offers the ability to slow down discourse. After all, as Paul Virilio reminds us: "the field of freedom shrinks with speed"; we find our options limited when we are forced to make decisions quickly.[8]

This question of speed obviously problematizes transhumanism's utopian potential; however, the problem both Stengers and Virilio identify in the cybercultural milieu that provides much of the exigence for transhumanism in the present is the lack of a space for critical deliberation, a space the idiot may be able to open up. As I have been making the case for in the prior two chapters, the longer history of evolutionary futurism contains an entire rich archive of these deliberative spaces. In this chapter I analyze this problem of critical reflection through the theme of *ease* in contemporary transhuman discourse. While transhumanism has a robust understanding of the technological problems involved in shaping a radically altered future for humanity, the social and cultural dimensions of such a problem are less well considered. To exemplify this problem, consider the science fiction writer Charles Stross, one of transhumanism's most effective and persuasive critics, who dramatizes a transhuman intelligence explosion in *Singularity Sky* (2003) as a dark parody of the chaos of the Russian Revolution in 1917, including the presence of a variety of extropian soviets and a Marxist–Gilderist dogma-spouting party functionary, Burya Rubenstein. In the book, Rubenstein meets the collective-intelligent Critics, a group that follows these transhuman upheavals and offers guidance (the Critics appear to be a collection of bootstrapped swarm intelligences made up of the uploaded minds of rabbits, mole rats, and other animals). Discussing revolutionary ideology and practice with Burya, one Critic says that of all the worlds they visit, "None of you people *ask* anything . . . Food, yes. Guns, yes. Wisdom? No."[9] The conversation goes on to reveal that the Critics stockpile a variety of philosophical and economic schemes that can help suddenly posthuman consciousnesses navigate the rigors of their new world, but nobody ever asks for that help. To dramatize this, Burya later meets a citizen of his world who has fashioned himself a new body that is half rabbit, half battle tank, bristling with weapons systems and armor. As Burya observes, "Many of the former revolutionaries had gone overboard on the personal augmentations offered by Festival, without realizing that it was necessary to modify the central nervous system in order to run them. This led to a certain degree of confusion."[10] As Stross draws out in this scene, the transhuman explosion in *Singularity Sky* fails (as they almost always do in his novels) because humans are consistently unaware of the degree of difficulty involved in radically altering humans in a world with no economic scarcity or death.

To further emphasize the difficulty of evolutionary futurism in the face of an overwhelming cultural obsession with ease, I juxtapose one thread of contemporary transhumanism, called "abolitionism,"[11] with the work of Pierre Teilhard de Chardin, a Jesuit mystic and paleontologist who first coined the modern usage of "transhuman" and laid much of the philosophical groundwork for evolutionary futurism. In abolitionism, which specifically focuses on the technoscientific elimination of human suffering, the idea of a future without suffering is an easily attainable pathway to a radical posthuman existence. This position, I argue, is emblematic of the lack of critical deliberation and focus on speed that aligns transhumanism with the broader entrepreneurial culture that Stengers wants to resist with the idiot. Teilhard, on the other hand, specifically seeks a path beyond the human through suffering. In juxtaposing these two positions, I show how Teilhard, though rejected for being too mystical and too religious by many contemporary transhumanists, offers a more rational approach to a scientific, utopian, evolutionary futurist practice.

On Beyond Suffering: Abolitionism and *The Hedonistic Imperative*

In *Man into Superman*, Robert Ettinger, the inventor of cryonics and one of the founding thinkers of transhumanism, asks, as we saw in the Introduction, why it is that some, given the opportunity to live forever, would choose instead to die. Moving on from those who "will not concede . . . it is better to live than to die," he offers the following image:

> For those who can enjoy leisure in traditional lazy and frivolous ways, a world of delight opens wide: an open-ended future may mean a month wandering the Canadian wilderness, a winter basking on Pacific beaches, a year listening to Bach and Mozart, or Simon and Garfunkel. La dolce vita can become cloying or even disgusting, but some things are likely to wear well: soft grass, a fresh breeze, fleecy skies, a cool drink, crisp snow, a warm hand, a familiar voice—can a thousand years of these be too much?[12]

This vision of the good life forever is not only compelling in selling Ettinger's postdeath vision of immortality as endless leisure; it inaugurates an important trope in the development of a properly transhuman rhetoric. From Ettinger's vision of immortality as a key component of the transhuman project there emerges an entire discourse in transhumanism organized around the end of suffering and the creation of new horizons for humans built specifically on limitless happiness. This model of a postsuffering transhumanism is called "abolitionism," a field organized around the manifesto, titled *The Hedonistic Imperative*, authored by utilitarian philosopher David Pearce. Pearce's manifesto begins,

> This manifesto combines far-fetched utopian advocacy with cold-headed scientific prediction. The Hedonistic Imperative outlines how nanotechnology and genetic engineering will eliminate aversive experience from the living world. Over the next thousand years or so, the biological substrates of

> suffering will be eradicated completely. "Physical" and "mental" pain alike are destined to disappear into evolutionary history. The biochemistry of everyday discontents will be genetically phased out too. Malaise will be replaced by the biochemistry of bliss. Matter and energy will be sculpted into life-loving super-beings animated by gradients of well-being.[13]

Pearce follows a number of tropes associated with evolutionary futurism, and he especially stresses the idea of freeing humanity from evolutionary legacy, arguing that in the future "the deliberate re-creation of today's state-spectrum of normal waking and dreaming consciousness may be outlawed as cruel and immoral."[14] We must, Pearce advocates, free "ourselves from the nightmarish legacy of our evolutionary past" and move into states of happiness heretofore unimagined.[15]

Like the religious impulse Max More seeks to replace with transhumanism, Pearce views our "mood-congruent pathologies of the primordial Darwinian psyche" as part of a primitive and outmoded biological system of motivation,[16] writing that

> the metabolic pathways of pain and malaise evolved only because they served the inclusive fitness of our genes in the ancestral environment. Their ugliness can be replaced by a new motivational system based entirely on gradients of well-being.[17]

Pearce is quick to stress that this vision of a postsuffering future is not the same as the false or chemically blissed out image of certain kinds of drug addicts, explaining that his vision of humanity in an abolitionist, hedonistic future is *not* like an experimental rat with a wired-up "mesolimbic dopamine system" who becomes an addict, going on "frenzied bouts of intra-cranial self-stimulation for days on end."[18] This image of the wired rat is "unedifying to all but the most unabashed hedonist," and yet, Pearce is not quite ready to be done with it: "more subtly engineered human counterparts of the euphoric rat are perfectly feasible. Centuries hence, any pleasure-maximising ecstatics will be using their personal freedom to exercise what is, in an ethical utilitarian sense, a legitimate lifestyle choice."[19] Beyond viewing intense, engineered pleasure as a reasonable future life choice, Pearce wants to suggest that, unlike the wired-up rat endlessly stimulating its dopamine reward center, the applications of this abolitionist hedonism will serve as a new, transhuman goal structure:

> Many dopamine-driven states of euphoria can actually enhance motivated, goal-directed behaviour in general. Enhanced dopamine function

> makes one's motivation to act stronger, not weaker. Hyper-dopaminergic states tend also to increase the range of activities an organism finds worth pursuing.[20]

From this idea, Pearce proposes to build a new "pain-free post-Darwinian Era" in which hedonism becomes "the moral foundation of any future civilisation" using a process he calls "paradise-engineering" to shape a heaven on Earth through genetic manipulation, nanotechnology, and cybernetically rewiring the brain.[21]

For Pearce, this paradise-engineering is an entirely rational and scientifically plausible program of creation. In a 2010 piece for *h+ Magazine,* titled "Top Five Reasons Transhumanism Can Eliminate Suffering," he points to emerging research in genetic engineering as a means for removing human suffering:

> We'll shortly be able to choose the genetically-shaped pain thresholds of our future children. Autosomal gene therapy will allow adults to follow suit. Clearly, our emotional response to raw pain is modulated by the products of other genes. But recent research suggests that variants of the SCN9A gene hold the master key. Thus in a decade or two, preimplantation diagnosis should allow responsible prospective parents to choose which of the SCN9A alleles they want for their future children—leading in turn to severe selection pressure against the SCN9A gene's nastier variants.[22]

Pearce's position on this issue is very interesting. As he makes clear in "The Abolitionist Project," a talk given at Oxford in 2007, there is a spectrum of suffering at work in his thought that cuts across statements about genes, as that paper focuses on the similarities and differences between raw pain (governed by SCN9A) and psychological pain (which blackens the mood and could drive people to suicide). For Pearce, both "agony after catching your hand in the door" and being "distraught after an unhappy love affair" can be removed through paradise-engineering. However, differentiating some of the potentially suspect remarks about finding subtler versions of the wired-up, onanistic rat, he helpfully suggests that the goal of abolitionism is "recalibrating the hedonic treadmill rather than simply seeking to maximize superhappiness."[23] In this way, I read Pearce as suggesting not only that we can opt out of the pain of physical existence but also that we can engineer a world free from, for instance, the little voice in the back of our heads that tells us we are not good enough. From this perspective, Pearce's explanation that hedonism provides extended and perhaps even more solid motivation—for activity, for exploration, and for the creation of newer and better understanding—seems compelling.

Embodied Suffering: Lyotard on Extropian Transhumanism

However, Jean-François Lyotard's "Can Thought Go On without a Body?" offers an account of suffering and pain very different from the one on offer in *The Hedonistic Imperative,* and, more important, like Stenger's idiot, Lyotard's account of philosophy and suffering stands to disrupt the enthusiasm of Pearce's abolitionist project, that, as Stengers figures it, is often "deaf and blind to the question of the world that [entrepreneurs'] efforts contribute towards constructing."[24] Lyotard's essay—the most extended and direct engagement with the extropian terrain of transhumanism in poststructural thought—stages a dialogue between an advocate of postembodiment extropianism labeled "HE" and a philosopher of the body labeled "SHE."

Over the course of the dialogue, which hinges on the inevitable heat death of the universe, HE defends an extropian position: uploading minds into computers may prove to be a way to escape the fact that, at the end of things, the sun will go out and life will be over. As HE stresses, the fact that this death of the sun is inevitable is especially a problem for a post-Kantian philosophy: "with the disappearance of earth, thought will have stopped—leaving that disappearance absolutely unthought of. It's the horizon itself that will be abolished."[25] There will be no thought of this solar death because there will be no one there to think it. This death is a death beyond death; it is a death that cannot be mourned: "All the events and disasters we're familiar with and try to think of will end up as no more than pale simulacra."[26]

However, HE has the answer: if we upload our bodies into computers, we will not be obliterated when the sun dies. Instead, we can engineer a postembodiment paradise that will sustain us as digital constructs. For HE, the solution is simply a technological one,

> Now: the hardware will be consumed in the solar explosion taking philosophical thought with it (along with all other thought) as it goes up in flames.
>
> So the problem of the technological sciences can be stated as: how to provide this software with a hardware that is independent of the conditions of life on earth.
>
> That is: how to make thought without a body possible.[27]

In order for thought to continue after the absolute death of the sun, Lyotard's scientifically minded HE repeats a commonplace transhuman argument: "we can shift the software of the human brain onto different hardware." Importantly, HE also proleptically dismisses philosophical objections to this idea by arguing that "your philosophy is possible only because the material ensemble called 'man' is endowed with very sophisticated software."[28] This argument for mind uploading, popularized by Hans Moravec in *Mind Children* (1990) and Raymond Kurzweil in *The Singularity Is Near* (2006), views the human as a computer and actualizes the mind/body split popular in humanist philosophy since Descartes. In this evolutionary futurist version of the mind/body dualism, however, the brain is a sophisticated computer program running on a particularly faulty meat computer. This logic is also, importantly, underscoring Pearce's work on abolitionism: if we change or remove negative inputs to this software object (the mind), we can create more favorable, desirable, pleasurable outputs. This idea, more than a metaphor—that the human is a computer and the mind is software—is core to certain contemporary transhuman notions of biohacking, quantified selfhood, and other approaches that seek to optimize the algorithms of consciousness. In summarizing his answer to the essay's titular question, HE suggests, "'without a body' in this exact sense: without the complex living terrestrial organism known as the human body. Not without hardware, obviously."[29]

SHE, who represents the philosophical objections to HE's rational, scientific account of uploading, begins to dismantle this position by referencing the terms we saw deployed by Pearce. Specifically, SHE's argument against the extropian position hinges on the incompleteness of simulation. Our human minds work by analogy and, SHE argues, this method "never satisfies the logical demand for complete description. In any serious discussion of analogy it's this experience that is meant, this blur, this uncertainty, this faith in the inexhaustibility of the perceivable."[30] Writing also functions like this for SHE: "when it stops . . . it's only suspending its exploration for a moment (a moment that might last a lifetime) and that there remains, beyond the writing that has stopped, an infinity of words, phrases, and meanings in a latent state."[31] SHE goes on to claim that, for uploading to work and for these uploaded minds to be actual re-creations of human beings, not mere simulations,[32] "real 'analogy' requires a thinking or representing machine to be *in* its data *just as* the eye is in the visual field or writing is in language. It isn't enough for these machines to simulate the results of vision or of writing fairly well."[33] Only when *in* data will an upload or an AI ever be anything more than "a poor binarized ghost of what it was beforehand."[34] Thought, in SHE's account, depends on a body in order to do this, "it's a matter . . . of 'giving body' to the artificial thought."[35]

For SHE, embodiment is central to thought because the world itself, the reality we move through in our day-to-day lives, is the stuff from which we form our knowledge. To extrapolate on Lyotard's position by way of an explanation, we can consider the abstract Turing machine, the theoretical model of a computer developed from Alan Turing's work on computation. A Turing machine, as described by Turing, consists of an infinite paper tape that contains a series of marks in individual cells and a read/write head that moves along the tape reading and writing data in a manner that is dictated by the instructions being read off and written to the tape. This theoretical machine can serve to model any computational machine and describes a whole set of problems that are solvable by computation. For the purposes of explaining SHE's statement of thought being "*in* its data," we can note that in a Turing machine, the read/write head and the apparatus that moves that head are outside and separate from the tape that constitutes its data "world." This model of computation positions thought as somehow separate from the world (continuing the tradition of Cartesian mind/body dualism) and is perfectly fine for understanding computers, but a growing body of feminist philosophy and emerging cognitive science research suggests that this critical distance from our data is not how human thought works. Instead, SHE's explanation of being "in" data would look something more like Alfred North Whitehead's process philosophy, in which the human sensorium is constantly bombarded by an undifferentiated and often overwhelming flux of the world. From this flux, our perception is able to extract and reflect on meaningful forms. As SHE argues,

> if you think you're describing thought when you describe a selecting and tabulating of data, you're silencing the truth. Because data aren't given, but givable, and selection isn't choice. Thinking, like writing or painting, is almost no more than letting a givable come toward you.[36]

In this way, we are in the world, in our data. The flux of being passes through us, and our perceptual apparatuses make cuts that become our experience of a much richer world.

Further, this cutting is what Lyotard means by thinking analogically about the brain: our experience of reality is only ever the cut-up version of a larger flux that we experience. In other words, we only ever experience an analogy to reality, not reality itself; we leave things out. Not understanding this left-out portion, SHE suggests, means we risk inaccurately simulating ourselves when we upload. For SHE,

> the mind isn't "directed" but suspended. You don't give it rule. You teach it to receive. You don't clear the ground to build unobstructed: you make a little clearing where the penumbra of an almost-given will be able to enter and modify its contour . . . This kind of thinking has little to do with combining symbols in accordance with a set of rules.[37]

Thus, for SHE, the impossibility of mind uploading lies in a flawed understanding of human thought. We are not rule driven when we think; instead, as she said, we clear out spaces where things don't make sense in order to fill them with these giveables from our data world.

Moreover, SHE goes on to argue that our cut-up experience of reality and the limits analogic thought imposes on our consciousness constitute what SHE calls "the pain of the unthought" and stands as the central motivating complex in SHE's understanding of human consciousness. These "almost-givens" pain us, because we do not understand them. SHE asks "will your thinking-, your representing-machines suffer? What will be their future if they are just memories?"[38] In answer to this question, SHE reminds HE,

> the unthought hurts because we're comfortable in what's already thought. And thinking, which is accepting this discomfort, is also, to put it bluntly, an attempt to have done with it. That's the hope sustaining all writing . . . that at the end, things will be better. As there is no end, this hope is illusory. So: the unthought would have to make your machines uncomfortable, the uninscribed that remains to be inscribed would have to make their memory suffer . . . Otherwise why would they ever *start* thinking? We need machines that suffer from the burden of their memory.[39]

This idea of pain is different from the versions of psychic or physical pain that Pearce seeks to engineer out of hedonistic posthumans, but it is still pain. For Pearce, the meaning of pain is always negative; through SHE, Lyotard articulates the idea that suffering can also, sometimes, be generative. This is the question SHE asks of extropian mind uploading, but also, I think, of abolitionism: what if the psychic suffering of something else we do not understand is core to our motivational system? How do we abolish pain or risk experimenting with the abolition of all suffering, if this suffering is core to our ability to think? I find SHE's account here to be persuasive, but I also do not think it entirely invalidates Pearce's program but rather dramatically complicates it. Pearce's idea of removing the pain that distracts us from unlocking more focused, more fully realized versions of ourselves is a lofty and, I think, laudable one; however, SHE's account of the pain of the unthought provides important insight into the role that certain kinds of suffering—especially the suffering we experience at not being able to understand some aspect of our data world—play in the constitution of an ennobling human existence.

To resolve this bind, I want to look, following Stengers, to another idiot from the history of philosophy. At the end of "On Truth and Lying in a Non-moral Sense," Friedrich Nietzsche juxtaposes two conceptual personae, the intuitive man and the rational man. Nietzsche first describes the intuitive man (a conceptual persona that, I would argue, has much in common with Stengers's idiot):

> The man of intuition, standing in the midst of a culture, reaps directly from his intuitions not just protection from harm but also a constant stream of brightness, a lightening of the spirit, redemption, and release. Of course, *when* he suffers, he suffers more severely; indeed he suffers more frequently because he does not know how to learn from experience and keeps on falling into the very same trap time after time. When he is suffering he is just as unreasonable as he is when happy, he shouts out loudly and knows no solace.[40]

The intuitive man, who stands "in the midst of a culture" and "reaps directly," is a possible inspiration for SHE's evocation of one who is in the data of its lifeworld. However, as Nietzsche suggests, the intuitive man also more directly experiences the suffering of the unthought, given that intuition, being in a data world, does not safeguard through an array of conceptual and explicative defense strategies. On the other hand, Nietzsche presents the rational man:

> How differently the same misfortune is endured by the stoic who has learned from experience and who governs himself by means of concepts! This man, who otherwise seeks only honesty, truth, freedom from illusions, and protection from the onslaughts of things which might distract him, now performs, in the midst of misfortune, a masterpiece of pretense, just as the other did in the midst of happiness: he does not wear a twitching, mobile, human face, but rather a mask, as it were, with its features in dignified equilibrium; he does not shout, nor does he even change the tone of his voice. If a veritable

> storm-cloud empties itself on his head, he wraps himself in his cloak and slowly walks away from under it.[41]

In contrast to the intuitive man, the rational man stoically brushes off all suffering and performs an even-keeled, stiff-upper-lip approach to suffering. In these two conceptual personae, we see the range of possible human responses to being in our data world: the pretense of hedonism and absolute suffering and the pretense of rational detachment and noble, silent suffering. I fear Pearce's approach to hedonism veers too closely to the intuitive man, one who is unable to parse the nature of the world and views pain only as negative input into the human biocomputer. For the remainder of this chapter, I focus on building a theory of evolutionary futurist Utopian overcoming that charts a more even course between the two pools—rational and intuitive—introduced by Nietzsche. In the Utopian science of Pierre Teilhard de Chardin, we find a way to balance these two factors, striking a balance that offers both the hedonism of Pearce and SHE's concept of the pain of the unthought.

The Amazing Disappearing Jesuit: Teilhard in the History of Transhumanism

While in "Can Thought Go On without a Body?" Lyotard is clearly engaging with some aspect of transhumanism, HE's position regarding the use of extropian mind uploading to escape from solar death is in need of an accurate citation, as Wikipedia would put it. One possible source of HE's specific position—rather than the general extropian ethos against which SHE's rebuttal serves—might be Frank J. Tipler's singular transhuman work, *The Physics of Immortality*. Tipler—a transhumanist, physicist, and born-again Christian—explores in *Physics* the possibility of escaping from the heat death of the universe while also building a cosmic, computerized version of God in the future, all thanks to a complex (though scientifically rigorous) scheme to colonize the universe with intelligent machines. Tipler's book, at its core, is an attempt to use evolutionary futurist rhetoric and cutting-edge technoscience to build a version of Christian eschatology, which he grounds in scientific methods instead of a mystical style he finds unacceptable. Point by point, Tipler addresses various aspects of the second coming of Christ through his vision of an extropian universe. Humans (or our posthuman descendents) will spread human intelligence throughout the universe using von Neumann probes and be able to prevent the heat death of the Universe. Originally proposed by John von Neumann, these probes are a form of self-replicating machine that are designed to be shot into space like pellets from a shotgun. On encountering a stellar body with sufficient resources, the probe would replicate itself and scatter more probes. If the stellar body is capable of supporting human life, the remaining probe would initiate terraforming, thus using randomness to seed the universe with human life. In a table labeled "Important Events in Future History," Tipler outlines the various milestones on the universe's path to a seemingly inevitable heat death (such as "Sun expands to engulf the Earth" and "Dead planets are detached from dead stars via stellar collisions").[42] However, he comments on this timeline that, in contrast to Lyotard's account of heat death, "this table assumes that life will not interfere with the evolution of matter. In reality, of course, life will. For example, rather than let Earth be vaporized seven billion years from now, our descendants would have long before taken the entire planet apart in order to use the material to expand the biosphere."[43] To radically simplify Tipler's point, intelligence is an important, and often overlooked, counterforce to entropy in the interstellar physics of the future.[44]

Eventually, in Tipler's evolutionary futurist eschatology, the network of von Neumann probes will represent a universe-spanning computer that, as one consequence, will be able to compute earlier states of matter from present data. For Tipler's purposes, this ability yields the possibility of reanimating the dead, as we will be able to compute prior states of matter (including human beings) and then simulate them within this massive computer. Moreover, importantly, Tipler's universe-spanning computer can impact the molecular behavior of the entire universe. As heat death arrives, this universal network can manipulate the shape of the universe as it cools:

> If, however, the universe stays the same size in one direction while it contracts in the other two directions, radiation in the latter directions will become hotter than in the stationary direction. This means that the directions of contraction will be hot spots in the sky and the other direction will be a cold spot. This temperature difference in different directions will power life in the far future just as the hot spot in the sky called the Sun powers life on Earth today.[45]

For Tipler, then, the ability of intelligence, represented by a literally God-like computer, will be able to check the ultimate and final death of the universe. This unequal collapse will allow our hypothetical posthuman offspring to live on indefinitely.

I mention Tipler's plan for two reasons: one, his account of intelligence at cosmic scales provides an interesting rebuttal to Lyotard's discussion in "Can Thought Go On without a Body?" While Lyotard has shaped an important account of the role of suffering in shaping our motives (a rebuttal that directly addresses the hedonism of Pearce's scheme), Tipler's version of an extropian eschatology shifts the register of suffering up a notch. The ultimate knowledge of our individual deaths, the small deaths that stand in contrast to the absolute solar death in Lyotard's account, is part of our motivation for the pain of the unthought, I would argue. By grappling with the absolute heat death of the universe (slated to happen, by the way, 10 trillion years *after* the solar death discussed by Lyotard), Tipler shifts the register of suffering: even if our species attains the kind of cosmic evolutionary expansions discussed in evolutionary futurism, the absolute heat death of the universe is a suffering our immortality must contend with. Put simply, while Lyotard's argument is capable of checking some aspects of a hedonistic version of transhumanism, his argument does not move up to the cosmic scale on which Tipler argues. Tipler's system of the future shifts beyond the solar system and thinks on a grander cosmic scale than many other transhuman thinkers. As we shall see, however, his focus on science and banishment of mysticism causes him to miss some of the key concepts important to a cosmic version of transhumanism that *is* grounded in the mystical.

I mention this cosmic version of transhumanism because *The Physics of Immortality* is an important updating and rewriting of Pierre Teilhard de Chardin's evolutionary futurist classic, *The Phenomenon of Man.* Teilhard was a French Jesuit paleontologist who, through a series of works suppressed during his life by the Catholic Church, articulated a system of cosmic intelligence that sought to square the discourse of evolutionary biology with Christian eschatology. While Tipler fills in the math, much of the original insight into a cosmic future humanity and the creation of a divine intelligence in that future is from Teilhard. Moreover, Teilhard's commitment to a mystical framework for his project and his use of Christian imagery and terminology to advance it more strongly suggests a Utopian social project than Tipler's rewriting, which is much more invested in the nuts and bolts of actualizing a divine superintelligence in the future. Using Fredric Jameson's distinction between Utopian Science and Utopian Ideology in *Archaeologies of the Future,* we can see Teilhard as providing the "Ideology" and Tipler providing a possible "Science" behind this Utopian scheme.

Teilhard's main contribution to evolutionary futurism is the concept he called noösphere. For Teilhard, human thought is becoming increasingly reflexive and capable of organizing itself at great distances as a result of emerging, global-scale technologies (to Teilhard, the airplane and the undersea telephone cable are emblematic of this planetary emergence). This ordering of thought itself into a new and dominant zone of expression for élan vital (Teilhard was heavily influenced by Henri Bergson's creative evolution[46]) was part of a longer general complexification that marks the evolution of matter on Earth. Given this long history, this concept of noösphere has a particularly complex lineage. As Marco Bischof summarizes in his article "Vernadsky's Noösphere and Slavophile Sobornost," noösphere emerges from a collaboration between Teilhard, the Russian geologist Vladimir Vernadsky, and French philosopher Édouard Le Roy (a student of Henri Bergson) in the years following World War I.[47] Vernadsky was in Paris on a lecture tour, popularizing his groundbreaking work on geology, when Le Roy and Teilhard, who had been collaborating on work designed to extend Bergson's theories into a theory of postnatural evolution, attended his lectures.[48] Vernadsky's work, which popularized the modern usage of "biosphere" and "geosphere," argues that geological evolution functions through a series of spheres of increasingly complex matter. When the geosphere was dominant, matter increasingly complexified through rock and metals until giving rise to the basic elements of life (single-celled organisms). At this point, the biosphere began to blanket the earth as the dominant configuration of matter (moving from mineral to life). As Vaclav Smil summarizes, in "one of his French lectures in 1925," Vernadsky identifies "humanity as a new geological, perhaps even cosmic, force resulting from human intelligence," which suggests that Vernadsky was thinking along lines similar to Le Roy and Teilhard, for whom thought was becoming a new evolutionary agent above life itself.[49] Within this context, Le Roy is purported to have suggested that humans, or more accurately our ability to think, may in fact constitute a new sphere of complexity.

Given that "noösphere" is a figure for collective intelligence, it is appropriate that each figure identified others as having been responsible for its creation. In a 1943 essay titled, "The Biosphere and the Noösphere," Vernadsky claimed that Édouard Le Roy, "in his lectures at the Collège de France in Paris, introduced in 1927 the concept of the noösphere as the stage through which the biosphere is now passing geologically. He emphasized that he arrived at such a notion in collaboration with his friend Teilhard de Chardin."[50] As Bischof summarizes, Vernadsky primarily used "noösphere" to refer to human reshaping of the Earth itself, thus representing our species' rise to a geological force, instead of the idea of a space of collective, shared personality as Teilhard came to use the term. However, Bischof's article is incredibly important for situating Vernadsky's usage of noösphere within the larger Slavophile and Russian Cosmist traditions that shaped his work on the biosphere. By tracing the concept of *sobornost,* Bischof shows that Vernadsky probably shared the cosmic implications for the noösphere found in Teilhard's writing.[51] However, of the three, Teilhard published most extensively on the topic (especially in its evolutionary futurist formation), which is why I focus on Teilhard's version in this chapter.

For Teilhard, the noösphere represents "a true pole of psychic convergence: a centre different from all the other centres which it 'super-centres' by assimilation: a personality distinct from all the personalities it perfects by uniting with them."[52] Just as the geosphere ultimately produced life through a process of increasing complexity, so is evolution creating, in us, a new and everywhere emergent force, as different as life is from the mineral, founded on thought just as the biosphere was organized around life and the geosphere organized around minerality. Teilhard's task becomes, from this fundamental insight, the process of describing what this shared space of personality might become. Essentially, he is trying to describe a space in which individuality may still exist but where the totality of all that exists will also be available to all, a kind of postindividual intelligence. Teilhard resorts to the long tradition of mystical writing, which often resorted to poetic rather than scientific or traditional philosophic rhetorical style in order to accomplish a difficult task: documenting a way of viewing self and world that is often at odds with the day-to-day experiences of consciousness. For Teilhard, the experience of the noösphere, though a scientifically valid concept in his mind, was also a mystical one that demanded a kind of hybrid scientific and poetic writing.[53]

This mysticism is a major problem for the reception of Teilhard by contemporary transhumanism, perhaps best documented in Nick Bostrom's "A History of Transhuman Thought." This intellectual history of transhumanism treats Teilhard's concept of the noösphere as an explicitly *antitranshuman* idea because of its association with mysticism:

> The singularity idea also comes in a somewhat different eschatological version, which traces its lineage to the writings of Pierre Teilhard de Chardin, a

paleontologist and Jesuit theologian who saw an evolutionary telos in the development of an encompassing noosphere [*sic*] (a global consciousness) . . . However, while these ideas might appeal to those who fancy a marriage between mysticism and science, they have not caught on either among transhumanists or the larger scientific community.[54]

Despite finding strong associations between Teilhard and the early origins of transhumanism, for Bostrom mysticism has no place in the philosophy of contemporary transhumanism. It was Teilhard's collaboration with Huxley that directly led to Huxley's early essay on the topic; additionally, as Richard Doyle reported finding amongst the Huxley papers at Rice University, a letter from Teilhard to Huxley contains what may be the first modern use of the term "transhuman." Bostrom's position exceeds just considerations of Teilhard; as a rule, Bostrom routes transhumanism around this question of mysticism in "A History of Transhuman Thought" by connecting it to the long history of humanism and specifically utilitarian philosophy. This move represents only one of the many attempts to ground transhumanism as a rational discourse about an irrational future in order to make the movement more rhetorically soluble to policy makers, scientists, and the general public.

However, Teilhard offers a lot of suggestive concepts for thinking about transhumanism as an evolutionarily astute form of Utopianism. As Bostrom suggests in the quote above, Teilhard was the first transhuman thinker to articulate the idea of a consciousness singularity, but he did so in a form radically different from Vernor Vinge's influential articulation. Vinge, in both his work as a computer science professor and as a professional science fiction writer, connects the rhetoric of the superman boom in science fiction with the proliferation of computing power on the surface of the Earth, best encapsulated by Moore's Law.[55] Based on I. J. Good's earlier concept of an imminent "intelligence explosion," the increase in computing power represented by Moore's Law suggests four possible human futures that Vinge enumerates in "The Coming Technological Singularity":

- Computers that are "awake" and superhumanly intelligent may be developed. (To date, there has been much controversy as to whether we can create human equivalence in a machine. But if the answer is "yes," then there is little doubt that more intelligent beings can be constructed shortly thereafter.)

- Large computer networks (and their associated users) may "wake up" as superhumanly intelligent entities.
- Computer–human interfaces may become so intimate that users may reasonably be considered superhumanly intelligent.
- Biological science may provide means to improve natural human intellect.[56]

Vinge suggests that once intelligence (artificial or otherwise) is increased past a certain point (a point he speculates will happen between 2005 and 2030), a change will occur and that change "will be a throwing-away of all the human rules, perhaps in the blink of an eye—an exponential runaway beyond any hope of control."[57] Vinge further suggests that universal intelligence can be graphed as an exponential function. Like all exponential functions, intelligence will eventually start to accelerate toward infinity. For Vinge, this will be the result of the fact that, at a certain point, superhuman intelligence will figure out how to quickly manufacture even more intelligent beings, and so on, and so on (the term "singularity" refers in mathematics to the moment at which an exponential function, when graphed, explodes to infinity. This "and so on, and so on" of intelligence is considered by Vinge and Kurzweil to be an exponential function). Once this happens, civilization will have reached what Vinge calls the Singularity, "a point where our old models must be discarded and a new reality rules."[58]

Vinge's understanding of the Singularity, perhaps more than any other concept in science fiction history, points to the increasing reality–SF interchange in which we live. First popularized in an op-ed piece in the SF magazine *Omni,* Vinge then used the concept as background for his novels *Marooned in Realtime* (1986) and *A Fire upon the Deep* (1992), before disseminating the hugely influential scientific essay "The Coming Technological Singularity: How to Survive in the Post-Human Era" over the Internet in 1993. This piece, even more than his novels, led to a wider acceptance and consideration of the Singularity, and the larger fate of the human race in an era always just on the cusp of intelligent machines. In fact, as an example of how ubiquitous the idea had become, Bill Joy, founder of enterprise computer manufacturer Sun Microsystems, penned his famous editorial in *Wired* magazine, "The Future Doesn't Need Us," in response to the ubiquity of pro-Singularity activists in Silicon Valley during the late 1990s and early 2000s—the same activism that would intensify into Raymond Kurzweil's Singularitarian movement in the 2010s.

Unlike Vinge's idea of technological and biological convergence, for Teilhard, a student of the evolution of mankind, the singularity became a means of articulating a unification of Christian eschatology and evolutionary biology in which neither discourse is made the dominant term. Teilhard, throughout his career (which is given an excellent overview in Julian Huxley's introduction to Teilhard's *The Phenomenon of Man*), articulated biological evolution as a force of increasing cognitive complexity that will ultimately result in a universal mind he called "noösphere." Teilhard's understanding of this emerging sphere, superseding the biosphere in which humanity evolved, was stated in terms of God: God ultimately stands to be created in the convergence of humanity within this new sphere of pure thought. This divinity leads to Bostrom's charges of mysticism. Bostrom continues about Teilhard:

> But the more general point that the transhumanist might make in this context is that we need to learn to think about "big-picture questions" without resorting to wishful thinking or mysticism. Big-picture questions, including ones about our place in the world and the long-term fate of intelligent life are part of transhumanism; however, these questions should be addressed in a sober, disinterested way, using critical reason and our best available scientific evidence.[59]

Essentially, then (and this is borne out in other essays in *JET* that discuss him), Teilhard's apparent "mysticism" invalidates his claims about the future of humanity. Bostrom's claim that "big-picture questions" "should be addressed in a sober disinterested way, using critical reason" implies, of course, that this is not the case with Teilhard's writing. To be fair, there is a poetic quality to Teilhard's writing, but this does not imply that his entire project is not "scientific." This is particularly salient because reason is often used as a justification for the increasingly rushed decisions demanded by globalized capital that Stengers seeks to resist through the figure of the idiot.

In reality, part of this disagreement regarding reason versus mysticism stems from a larger debate about philosophical style: Bertrand Russell once described Bergson's style (emulated by Teilhard in *The Phenomenon of Man*) in this way:

> Like advertisers, he relies upon picturesque and varied statement, and on apparent explanation of many obscure facts. Analogies and similes, especially, account for a very large part of the whole process by which he recommends his views to the reader.[60]

This rather dismissive stance (Russell frequently refers to Bergson's system as "irrationalism" in *The History of Western Philosophy*) itself derives from the overarching view among philosophers that Bergson, in a 1922 debate with Albert Einstein, failed to defend the intellectual turf of philosophy from an increasing encroachment by physics following the emergence of relativity.[61] Just as Russell attacks Bergson for having a lack of rigor due to his more poetic writing style, so Bostrom dismisses Teilhard because of perceived associations between poetic writing and mysticism.

Given the close associations that exist between contemporary transhumanism, science fiction, and theosophy, it makes sense that a rational utilitarian philosopher such as Bostrom would want to distance the movement today from the mystical trappings of Teilhard's work. However, if we believe that Teilhard is a mystic rather than merely resorting to a mystical style for its rhetorical affordances, we have to accept the popular opinion of his work, and, like so many other figures associated with the history of evolutionary futurism, ignore its contents. Teilhard, despite his interests in philosophy and theology, asserts in the very first sentence of the preface to *Phenomenon of Man* that "if this book is to be properly understood, it must be read not as a work on metaphysics, still less as a sort of theological essay, but purely and simply as a scientific treatise."[62] Teilhard—like Bergson, who felt that his book *Duration and Simultaneity* made serious contributions to the science of general relativity—positions *The Phenomenon of Man* as a work making serious contributions to the science of evolution, while merely implying revisions to theology and philosophy. Teilhard continues:

> In the first place, it deals with man *solely* as a phenomenon. The pages which follow do not attempt to give an explanation of the world, but only an introduction to such an explanation. Put quite simply, what I have tried to do is this; I have chosen man as the center, and around him I have tried to establish a coherent order between antecedents and consequents. I have not tried to discover a system of ontological and causal relations between the elements of the universe, but only an experimental law of recurrence which would express their successive appearance in time.[63]

In this way, Teilhard claims his work to be scientific: he is looking at humanity (or more broadly the organized intelligence instantiated by our species) as another scientific phenomenon. However, his work is difficult to absorb into the rational worldview projected by scientific rhetoric. As

Julian Huxley's introduction to *Phenomenon of Man* points out, Teilhard's work serves as a challenge to both religion and science:

> It is no longer possible to maintain that science and religion must operate in thought-tight compartments or concern separate sectors of life; they are both relevant to the whole of human existence. The religiously-minded can no longer turn their backs upon the natural world, or seek escape from its imperfections in a supernatural world; nor can the materialistically-minded deny importance to spiritual experience and religious feeling.[64]

Huxley's introduction is interesting in situating the response to Teilhard, especially within a more scientifically inclined framework. Huxley, whose life's work spanned C. P. Snow's two cultures (publishing both scientific works and humanistically minded social criticism), shows in his introduction that it is, while not impossible, very hard to read Teilhard as a synthesis of the *three* domains his work touches: science, philosophy, and theology.

While Teilhard asserts that his work is "purely and simply" a "scientific treatise," Huxley argues that the true import of the work is in its ability to provocatively combine science and religion for an evolving and technologically maturing humanity, similar to the position Max More hopes transhumanism can occupy in a postreligious, goal-oriented futurist society. In many ways, Huxley's introduction does a lot of damage to Teilhard's ethos and perhaps explains why, as Nick Bostrom points out, his ideas "have not caught on either among transhumanists or the larger scientific community." While Teilhard is emphatic about the scientific content of his work, Huxley's posthumous introduction suggests it is as much a work of religion and philosophy as of science, making it easier to avoid grappling with the implications of Teilhard's thought *as a science.* Having said that, however, taking Teilhard's assertions that the work of *Phenomenon of Man* is scientific, what claims can we make about the science of transhumanism?

Additionally, given the more rational, less poetic character of Teilhard's other works, such as "On Viewing a Cyclotron" and the essays compiled in *The Future of Man,* I think we *can* and *should* at least attempt to take Teilhard at his word regarding the scientific content of his work. Moreover, considering Teilhard's admittedly poetic, philosophical writing *as* science opens up some interesting means of considering what contemporary transhumanism is actually doing, as a science or a response to science. In fact, many of the rhetors involved in establishing contemporary transhumanism as a viable public field of discourse seek to legitimize their work through appeals to the seriousness of science (for instance, Bostrom's dismissal of Teilhard with the idea that transhuman "questions should be addressed in a sober, disinterested way, using critical reason and our best available scientific evidence").

To consider the "scienciness" of Teilhard, I instead want to look at a discussion of what science is and does from K. Eric Drexler's *Radical Abundance* (2013). In considering the difference between engineering and science, Drexler provides the following dichotomy:

> The essence of science is inquiry; the essence of engineering is design. Scientific inquiry expands the scope of human perception and understanding; engineering expands the scope of human plans and results.[65]

He then enumerates the following differences:

- Scientists seek unique, correct theories, and if several theories seem plausible, all but one must be wrong, while engineers seek options for working designs, and if several options will work, success is assured.
- Scientists seek theories that apply across the widest possible range (the Standard Model applied to everything), while engineers seek concepts well suited to particular domains (liquid-cooled nozzles for engines in liquid-fueled rockets).
- Scientists seek theories that make precise, hence brittle, predictions (like Newton's), while engineers seek designs that provide a robust margin of safety.
- In science, a single failed prediction can disprove a theory, no matter how many previous tests it has passed, while in engineering one successful design can validate a concept, no matter how many previous versions have failed.[66]

This typology helps in thinking about the scientific content of Teilhard. Considering the examples of contemporary transhuman practice documented in this book, how much does transhuman practice expand "the scope of human perception and understanding"? Instead, I argue, contemporary transhumanism often seeks more to expand "the scope of human plans and results," especially in the version Bostrom describes. In other words, contemporary transhumanism is often more a kind of posthuman, futurological engineering than an actual science, at least according to Drexler. By labeling transhumanism as "engineering," I do not mean to dismiss the contemporary discourse, merely to highlight the fact that many thinkers working on transhuman technologies are more interested in seeking specific solutions to the problems of the human body. Remember here that David Pearce, a cofounder of Humanity+ with Bostrom, has taken to referring to the shaping of a transhuman future as "paradise-engineering." Similarly, in his dismissal of Teilhard, Bostrom argues that transhumanism is interested in big problems from a "disinterested" perspective, not because of the new avenues for thought afforded by pondering philosophical issues like immortality, but in order to better understand "what policies it makes sense for humanity to pursue."[67]

The question then, returning to Teilhard, is whether his work still counts as science following Drexler's definition. While not making brittle or falsifiable claims about reality in *Phenomenon of Man,* Teilhard argues for an increased perception and understanding. After situating his work as science, Teilhard argues that his understanding of evolutionary futurism hinges on an expanded perception. His vision of the coming ultra-humanity, as he calls it, hinges on a union of minds in an emergent noösphere and, for him, "union increases only through an increase in consciousness, that is to say in vision."[68] As we shall see, Teilhard's idea of evolutionary futurism hinges on an expanded perception, a reorientation of consciousness toward the global scope made possible by telecommunications. How does one go about accomplishing this expanded vision, though? As Richard Doyle suggests in *Darwin's Pharmacy,* the introduction of chemical adjuncts such as LSD is one avenue (explaining why it is the general practice of a shamanic rhetoric). However, as Doyle further clarifies, these practices, that also often have mystical trappings, wreak havoc on narrow-minded understandings of scientific verifiability. The poetic language of *Phenomenon of Man,* then, can be embedded in the tradition of ecodelic rhetoric Doyle identifies: Teilhard's writing is a provocation to see the world differently, to emerge into a world-making consciousness, to become ultrahuman.[69] This kind of provocation is at odds with a common view of science, but how different is Teilhard's practice from the kind of paradoxical avant-gardism used to discuss general relativity and other advances in physics made during the twentieth century? We can begin to see that this important book enacts the kind of consciousness shift in its readers that Eric Drexler suggests is at the core of scientific practice. So despite being dismissed by contemporary "paradise-engineering," Teilhard's work constitutes an important approach to imagining an evolutionary futurist science and, as I explore in the next section, a Utopian one as well.

The Passion of the Ultrahuman

For Teilhard, the noösphere represents the moment in which "the specific orthogenesis of the primates (urging them toward increasing cerebralization) coincides with the axial orthogenesis of organised matter (urging all living things toward a higher consciousness)" to produce, primarily in humanity, the ability for consciousness to reflect on itself.[70] This involuted cognitive reflection constitutes a singularity for Teilhard: the transition from biosphere to noösphere. This rising to consciousness constitutes the human phenomenon for Teilhard.[71] Given this ongoing human phenomenon, the present mental state of humanity is akin to the first bacterium that eventually birthed the biosphere from the raw material of the geosphere. The noösphere currently exists but only in a natal sense.

For Teilhard, this rising complexification suggests an increasing global sense of unity. Similar to Vernadsky's wartime speculation that led to the view of "humanity as a new geological, perhaps even cosmic, force," Teilhard's experience as a stretcher bearer during World War I led him to experience a "fundamental vision . . . of plurality and the multitude, the multitude that surrounds us and the multitude that constitutes us, that is in restless motion around us, and that shelters within us."[72] Growing from an initial feeling of togetherness, this cosmic realignment of perception comes to stand at the core of Teilhard's vision. In the foreword to *The Phenomenon of Man,* Teilhard situates individuated consciousness as that which must be overcome on the path to a more complex noösphere. For Teilhard, our consciousness is a form of "bondage . . . It is tiresome and even humbling for the observer to be thus fettered, to be obliged to carry with him everywhere the centre of the landscape he is crossing."[73] In rare moments, however, "the subjective viewpoint coincides with the way things are distributed objectively, and perception reaches its apogee"; we experience the objective truth of the universe: we are one with it and not the center of it.[74] This claim to unity is perhaps the most contentious point in Teilhard's oeuvre. Like Bergson before him, Teilhard's assertion that "man is unable to see himself entirely unrelated to mankind, neither is he able to see mankind unrelated to life, nor life unrelated to the universe" leads to the accusation that Teilhard possesses a naive belief in animism, the ancient idea that all matter possesses a soul.[75] In addition to understanding mystical unity, Teilhard's writing here draws on contemporaneous advances in physics that suggest that observation is an act enmeshed with the phenomenon under observation; Teilhard is attempting to apply this insight to the study of humanity and not just to the study of atoms and particles. For Teilhard, just as an observer is involved in an atomic phenomenon by the act of observation, so are we always influencing and influenced by the emergent noösphere of which we are a part and that also exceeds us.

This reflection is born of science *and* mysticism but is propelled forward by emerging technologies:

> Through the discovery yesterday of the railway, the motor car and the aeroplane, the physical influence of each man, formerly restricted to a few miles, now extends to hundreds of leagues or more. Better still: thanks to the prodigious biological event represented by the discovery of electro-magnetic waves, each individual finds himself henceforth (actively and passively) simultaneously present, over land and sea, in every corner of the earth.[76]

This extension of the range of humanity, so potently prefiguring Marshall McLuhan's view of media as extensions of man, points to an increasing globality of consciousness for Teilhard. In his system, this increasing globality will ultimately yield to a fully conscious and reflective noösphere. He writes,

> The noösphere in fact *physically* requires for its maintenance and functioning, the existence in the universe of a true pole of psychic convergence: a centre different from all the other centres which it "super-centres" by assimilation: a personality distinct from all the personalities it perfects by uniting with them. The world would not function if there did not exist, somewhere ahead in time and space, "a cosmic point Omega" of total synthesis.[77]

This "cosmic point Omega" is the moment in the future that represents a fully realized consciousness singularity, the moment at which the universe fully becomes conscious of itself. This point of "total synthesis" which results in "a personality distinct from all the personalities it perfects by uniting with them" is perhaps one of the most challenging concepts in Teilhard's thought. Teilhard walks a very fine intellectual line, in his writing, around several of the Utopian forces of the twentieth century. He does not view Omega as a Communist loss of individuality in a sea of mass proletarians; nor does he see this Omega result in the kind of selfish hyperhumanity of libertarian transhumanism. He stresses that the noösphere is a space of communion with the whole that does not wholly subsume the individual. The difficulty of capturing such a supercentered consciousness, again, partly explains the opacity of his poetic language. The closest analog would be the experience of the personal God in Christianity, which gets back to Teilhard's lifelong goal of synthesizing theology and biology. Of course, in essence, this relationship between God and Omega is what ultimately led the Catholic Church to block publication of his work during his lifetime: Teilhard reverses and inverts Creation, changing its agent and its chronology by arguing for the creation of God by Man in the future.

The idea of a supercentered individual is one of the ideas that drops out of Teilhard's evolutionary futurism in contemporary transhumanism, which, however, does retain the idea of evolutionary technology, consciousness singularity, and human centrality. That said, without the central role supercentered personality plays in Teilhard's thought experiment, the whole system would seemingly collapse into the kind of hypertrophied egoism that dominates contemporary transhumanism. This deletion of the key Utopian insight of evolutionary futurism occurs even in the pages of Teilhard's own books, specifically in the introduction provided by Julian Huxley for *The Phenomenon of Man.* Based on their correspondence, Huxley would publish the essay "Transhumanism"—the essay that first suggested evolutionary futurism under its modern name. Huxley's essay, and especially his introduction to the book, is the first moment at which, I argue, the individual becomes the central unit of transhuman evolution.

Summarizing the contents of *The Phenomenon of Man,* Julian Huxley addresses Teilhard's idea of Omega, writing that "the combined result, according to Père Teilhard, will be the attainment of point Omega, where the noösphere will be intensely unified and will have achieved a 'hyperpersonal' organisation."[78] Huxley then makes a curious rhetorical move, especially for someone writing an introduction to a major work of thought: he admits he doesn't understand. Huxley writes:

> Here his thought is not fully clear to me. Sometimes he seems to equate this future hyperpersonal psychosocial organisation with an emergent Divinity: at one place, for instance, he speaks of the trend as a *Christogenesis*; and elsewhere he appears not to be guarding himself sufficiently against the dangers of personifying the non-personal elements of reality. Sometimes, too, he seems to envisage as desirable the merging of individual human variety in this new unity. Though many scientists may, as I do, find it impossible to follow him all the way in his gallant attempt to reconcile the supernatural elements in Christianity with the facts and implications of evolution, this in no way detracts from the positive value of his naturalistic general approach.[79]

This quotation marks, I argue, the beginning of the deviation from Teilhard by contemporary transhuman discourse. Huxley interprets Teilhard as making flawed arguments, by (ironically) pointing to the precise rhetorical moves that perform the core of his argument: the personification of nonpersonal elements of reality *and* the cosmic unity of all in a post-Omega noösphere. Huxley, in introducing the work, situates these key movements in Teilhard's thought as accidents or mistakes. Moreover, he dismissively chalks these mistakes up to the "supernatural" and non-scientific aspects of Teilhard's work. So in this supposedly supportive introduction, Huxley blatantly disregards the central challenge that Teilhard puts to readers of *The Phenomenon of Man*: to read a work couched in mysticism *as* a work of science. Where Lyotard reminds us that thought has to be embedded in its data, Huxley seeks to extract Teilhard's mystical language from his stated commitments to scientific enquiry. The work then becomes a kind of inspirational message of peace and love, instead of the serious performative challenge Teilhard thinks of it as.

In Huxley's reading of Teilhard, personality, attained by overcoming individuality, is the key to the advancement of human evolution. Huxley suggests that Teilhard's "concept of a hyperpersonal mode of organisation sprang from Père Teilhard's conviction of the supreme importance of personality."[80] He goes on to suggest that personality's "fuller achievement must be an essential aim for his evolutionary future."[81] However, finding where Teilhard discusses "personality" in *The Phenomenon of Man*, we first come across the following: "personality is seen as a specifically corpuscular and ephemeral property; a prison from which we must try to escape."[82] This statement would appear at odds with Huxley's summary in which the "attainment of personality" is the "essential element in man's past and present evolutionary success." For Teilhard, this statement *is* true, but not perhaps in the way that Huxley thinks. Teilhard is opposed to any view of human evolutionary futurism that partakes of a winner-takes-all model. For him, individual personality is as much a dead end as individualism itself. As he continues after calling personality a prison, "all our difficulties and repulsions as regards the opposition between the All and the Person would be dissipated if only we understood that, by structure, the noösphere (and more generally the world) represent a whole that is not only closed but also *centred*."[83] For Teilhard, this centering *is* personality. So, while Huxley seems to read Teilhard as calling for an intensification of individual personalities, Teilhard is actually arguing for a kind of universalization of personality. This universalization returns to the point relating the Christian God to Omega in his system: God is the only available means Teilhard has for articulating what a supercentered personality might look like. While Huxley may find problems with the "supernatural" elements of Teilhard, once again, we see that these elements are crucial to the whole project.

In Huxley's influential, and very Teilhardian, essay "Transhumanism," this misreading of Teilhard's usage intensifies. In reading through the essay, we can see the rhetorical movement from Teilhard's thought to contemporary transhumanism. Huxley, for instance, inaugurates the obsession with human potential: he defines humanity as "every man-jack of us begins as a mere speck of potentiality, a spherical and microscopic egg-cell."[84] Moreover, as we saw above, he continues to misunderstand the role of "personality" in Teilhard's model. For instance, when he continues to obsess over unlocking human potential, he writes:

> The great men of the past have given us glimpses of what is possible in the way of personality, of intellectual understanding, of spiritual achievement, of artistic creation . . . We need to explore and map the whole realm of human possibility, as the realm of physical geography has been explored and mapped. How to create new possibilities for ordinary living?[85]

Once again, in this passage, Huxley apprehends Teilhard's understanding of "personality" as a kind of refined, individual intellect (notably in contrast, at present, to "ordinary life"), rather than a means of grasping the hyperpersonal nature of an emergent universal mind. Moreover, though, observe in the above quote Huxley's rhetorical substitution: in the first sentence, "great men of the past" teach us "what is possible in the way of personality," but in the third sentence, "we need to explore and map the whole realm of human possibility."[86] For Huxley, then, we can conclude that "personality" and "potential" are the same thing (or at least inextricably linked).

Huxley's use of "the great men of the past" begins to point toward the vision that he sustains for the future of humanity. In further clarifying humanity's role in shaping an evolutionary rising, he writes:

> It is as if man had been suddenly appointed managing director of the biggest business of all, the business of evolution—appointed without being asked if he wanted it, and without proper warning and preparation. What is more, he can't refuse the job. Whether he wants to or not, whether he is conscious of what he is doing or not, he is in point of fact determining the future direction

of evolution on this earth. That is his inescapable destiny, and the sooner he realizes it and starts believing in it, the better for all concerned.[87]

The summoning of the image of the manager is important for understanding the return of the individual within Huxley's misreading of Teilhard. This distinction between Teilhard's understanding of evolutionary action and Huxley's usage of humanity being "appointed managing director" of evolution touches on a much longer rhetorical debate about the meaning of "control" in the era of cybernetics.

Timothy Leary's essay on cyberpunk culture, "The Cyber-punk: The Individual as Reality Pilot," handles this distinction in a way that helps to clarify Huxley's misinterpretation of Teilhard's claims. In Leary's essay, the figure of the cyberpunk—popularized by SF writers such as William Gibson and Bruce Sterling—becomes a model for being in the space of the postmodern. Leary suggests that the concept of the cyberpunk is actually transhistorical, suggesting that Prometheus was the first cyberpunk who hacked the prohibition against giving fire to humans.[88] For Leary, these cyberpunks, strongly individualist beings who have a deep understanding of technology, "were tolerated only at moments when innovation and change were necessary."[89] Leary's essay reconstitutes the cyberpunk by exploring translational slippage between Greek and Latin that underscores how we think about control in cybernetics. Reading Norbert Weiner's *Human Use of Human Beings,* Leary reminds his readers that "cybernetics" comes from the Greek word *kubernetes,* meaning "pilot." However, as Leary goes on to show, Weiner's use of this term to name his new science of control and communication actually draws on the Latin usage of the concept. In the Greek mindset, pilots, "sailing the seven seas without maps or navigational equipment, [were] forced to develop independence of thought."[90] As Leary continues, "the Athenian cyber-punk, the pilot, made his/her own navigational decisions."[91] In Roman usage, *kubernetes* becomes *gubernetes,* the root of "governor," which "means to control the actions or behavior of, to direct, to exercise sovereign authority, to regulate, to keep under, to restrain, to steer. This Roman concept is obviously very different from the original notion of 'pilot.'"[92] So, as Leary shows, there are actually two very different conceptions of ethos contained in the word "cybernetics." In the Greek ethos, the image Leary associates with William Gibson's hacker antiheroes, pilots are in charge of their own fate as they navigate between islands; in the Roman ethos, the vast empire is piloted from Rome through a series of bureaucratic innovations.

This distinction in the double meaning of "pilot" also functions to clarify the differences between Teilhard and Huxley. The Greek denotation of "pilot" is the one Teilhard most associated with his understanding of evolutionary overcoming. For Huxley, however, humanity is the "managing director" (which is of course the late-capitalist moment's version of a Roman governor) of evolutionary change. Huxley's transhumanism foregrounds inevitability and has much of the excitement of a corporate quarterly report: "how we did in bringing Cosmic Point Omega to our customers in Spring of 2014." While he begins his essay with the phrase that Teilhard also quotes favorably in *The Phenomenon of Man,* "as a result of a thousand million years of evolution, the universe is becoming conscious of itself," he continues by stating that "this cosmic self-awareness is being realized in one tiny fragment of the universe—in a few of us human beings."[93] In this way, Huxley foregrounds the managerial role of an elite in shaping the transhuman future of humanity.

In contrast to this elitism, in Teilhard's *The Phenomenon of Man,* the only mention of an elite in the entire book is consigned to a footnote appended to the statement that the doors of Omega "will open only to an advance of all together, in a direction in which all together can join and find completion in a spiritual renovation of the earth."[94] Teilhard attaches a footnote to "all together" that reads "even if they do so only under the influence of a few, an *élite.*"[95] In fact, in the original French, this idea of an "*élite*" is even more subordinate. Teilhard's original footnote reads "Fût-ce sous l'influence et la conduite de quelques-uns (d'une «élite») seulement." So, in Teilhard's French manuscript, the elitism is within quotes, in a parenthetical statement, within a footnote. In this grammatical distancing in Teilhard and rhetorical foregrounding in Huxley, we can see that the role of a benevolent elite who knows what's best for humanity, marginalized in Teilhard, becomes a central concept in Huxley's transhumanism. This textual shift from footnote to thesis is central to understanding the emergence of the ego-driven, hyperindividualist model of transhumanism that functions in the hedonism of Pearce or the neoliberal Singularitarianism of Raymond Kurzweil.

Moreover, this shift is tied to Huxley's main critique of Teilhard's thought: the idea that the noösphere emerges through a process of Cosmic "christogenesis" in *The Phenomenon of Man.* This concept, central to Teilhard's intellectual trajectory but very much dampened in his later work, shapes, as we have already seen, much of his understanding of a hyperpersonal, transhuman universe. Beyond being a sticking point for scientists *and* transhumanists, as Bostrom established, this idea is also central to Teilhard's lifelong struggle to get the Church's authorization to publish any of his writings. In Teilhard's system, the destiny of mankind is to actualize God in the future. Of course, as alluded to above, the potentially blasphemous corollary to this idea is that God does not yet exist in the present (though Frank J. Tipler in *Physics of Immortality* will suggest that what we experience as God are emanations from the future-created universal supercomputer). As Teilhard's personal God is inextricable from his concept of Cosmic Point Omega, so too is the suffering of Christ, as depicted in the Bible, integral to his conceptualization of the transhuman. Despite this centrality, one of the temptations is to read Teilhard's intellectual career as a gradual secularization. For instance, in *The Cosmic Life* (1916), his first major work of evolutionary futurism, Teilhard uses the word "God" 26 times in a 63-page work. By comparison, in *The Phenomenon of Man* he mentions "God" only 11 times in 319 pages. This shift in word usage from once every 2.4 pages to once every 29 pages would suggest a gradual deemphasis of Christianity in favor of more secular pursuits. However, the final paragraph of *The Phenomenon of Man* (which powerfully encapsulates Teilhard's entire approach) reads, "In one manner or the other it still remains true that, even in the view of the mere biologist, the human epic resembles nothing so much as a way of the Cross."[96] Reading this after a discussion of the philosophy of evil within the science of evolution is shocking and a fascinating conclusion to a major work, especially one that stresses optimism as its overarching ethos. More important for the issue at hand, however, this statement emphasizes the importance of suffering in Teilhard's system of transhumanism. In Teilhard, the suffering of Christ on the way to and during his crucifixion is the best way to think about the work of an evolutionary vanguard, just as the relationship between human and God is the best means for articulating the experience of noösphere that Teilhard has access to.

In "The Cosmic Life"—which the editors of *Writings in the Time of War* (in which it is collected) call his pre-Jesuit "intellectual testament"—Teilhard offers a strong case for thinking about evolutionary unfolding within the confines of Christ's suffering.[97] He suggests that "the truth of our position in this world is that *in it we are on a Cross*."[98] For Teilhard, this realization develops from an argument about, as we saw above, a cosmic shift in vision. The realization of the oneness of being shapes his evolutionary futurist project and calls out for a specific kind of action. He writes of the cosmically enlightened human:

> If he is to act in conformity with his new ideal, the man who has determined to admit love of the world and its cares into *his* interior life finds that he has to accept *a supreme renunciation.* He has sworn to seek for himself outside himself. He will now have to realize what this noble ambition will cost him.[99]

This cost is, of course, the suffering of Christ. In accepting the cosmic life,

> He may perhaps have to accept the role of the imperceptible atom which loyally, but without honour, carries out the obscure function for which it exists, to serve the well-being and balance of the Whole. He must agree to be, some day, the fragment of steel on the surface of the blade that flies off as soon as a blow is struck, the solider in the first wave of the attack, *the outer surface, made use of and sacrificed, of the cosmos in activity.*[100]

This idea of instrumentalization, of being "the surface of the blade that flies off as soon as a blow is struck," is very different from the managing director of evolution that Huxley sees as the role of the elite in actualizing transhumanism. Suffering, here, becomes an important way of thinking through the role of a transhuman vanguard in Teilhard's model. While sometimes the way of the noösphere means getting bionic arms and enhanced brains, sometimes it just means sacrifice.

This focus on suffering and sacrifice is unexpected given that Teilhard's work oscillates between the violent passion of Christ's suffering on the cross and a poetic love of the world. While Teilhard has many quotes about the power of love—including the most famous one "it is not a *tête-à-tête* or a *corps-à-corps* that we need; it is a heart to heart . . . if the synthesis of the Spirit is to be brought about in its entirety . . . it can only . . . be realized in a universal, mutual love," which is often anthologized in Christian self-help guidebooks with titles such as *Your Sacred Self: Making the Decision to Be Free* and *The Proof: A 40-Day Program for Embodying Oneness.* Teilhard's use of the term is less the self-empowering love of Elizabeth Gilbert in *Eat, Pray, Love* and more the self-sacrificing love described in John 3:16 ("For God so loved the world, that he gave his only begotten Son, that whosoever believeth in him should not perish, but have everlasting life").[101] For Teilhard, the love of the world is inherently routed through a need to suffer, a love that compels sacrifice and pain. To love the world is to be willing to suffer for it. Not exactly the stuff of self-help manuals.

For Teilhard, however, Christ is not merely the figure of individual suffering. In *The Cosmic Life,* Christ is discussed as "The Cosmic Christ," a much more multitudinous figure. Teilhard writes:

> Jesus Christ is united to all sanctified souls, and since the bonds that link souls to him in one single hallowed mass end in Him and meet in Him . . . Souls, however, are not a group of isolated monads. As the "cosmic view" specifically shows us, they make up one single whole with the universe, consolidated by life and matter. Christ, therefore, cannot confine his body to some periphery drawn within things; though he came primarily, and in fact exclusively, for souls, he could bring them together and give them life only by assuming and animating, with them, all the rest of the world; through his Incarnation he entered not only into mankind but also into the universe that bears mankind—and this he did, not simply in the capacity of an element associated with it, but with the dignity and function of directive principle, of centre upon which every form of love and every affinity converge . . . Christ has a *cosmic Body* that extends throughout the whole universe.[102]

For Teilhard, then, the figure of the Cosmic Christ is the unifying principle of the entire cosmos, an early version of the single pole that structures the move toward Cosmic Point Omega in *The Phenomenon of Man.* One upshot of this model of the universe is that Teilhard can claim that "by the incarnation, which redeemed man, the very Becoming of the Universe, too, has been transformed. Christ is the term *of even the natural* evolution of living beings; evolution is holy."[103] This claim, as we have seen, is central to Teilhard's entire project.

Therefore, the idea of the Cosmic Christ is important to Teilhard, but it is equally important for evaluating the contemporary transhumanism that mutates from his work. Specifically, the language and imagery Teilhard generates by associating evolutionary futurism with the Way of the Cross focuses on the suffering of Christ as a vehicle for thinking about the loneliness and pain involved in being of an evolutionary vanguard. This issue of suffering crucially complicates a scheme like Pearce's paradise-engineering, in which the future is marked by an unending increase of pleasure. As we have seen in this chapter, there is a significant body of evolutionary futurist thought—from Stengers to Lyotard and most prominently in Teilhard—that foregrounds the role that suffering plays in shaping thought, life, love, and Utopia itself. For Teilhard, the role of a vanguard, deemphasized as it is in *Phenomenon of Man,* is more to suffer for the future than to be early adopters of new tech trends or cutting-edge life hacks.

In this way, I think we can see in passing a bit of the character of Stengers's idiot in Teilhard's Utopian version of evolutionary futurism. By deemphasizing the individual and the vanguard, his concern lies with creating a version of the noösphere into which all may pass as equals while at the same time never forgetting the importance that the individual plays in shaping the richness of the whole. Here, Teilhard plays the role Stengers ascribes to the idiot: the one who says "slow down." As Stengers says in introducing the cosmopolitical vision in which the idiot plays such an important role, she writes that

> In the term cosmopolitical, cosmos refers to the unknown constituted by these multiple, divergent worlds, and to the articulations of which they could eventually be capable, as opposed to the temptation of a peace intended to be final, ecumenical: a transcendent peace with the power to ask anything that diverges to recognize itself as a purely individual expression of what constitutes the point of convergence of all.[104]

Stengers describes her project using vocabulary drawn from *The Phenomenon of Man*, especially the last utterance about the "purely individual expression of what constitutes the point of convergence of all": this is how Teilhard describes Cosmic Point Omega. In this way, we can argue that Stengers's call to slow down without reason why, the uncomprehending suffering of Nietzsche's intuitive man, and Teilhard's "surface of the blade that flies off as soon as a blow is struck" all share an evolutionary futurist lineage in which things are never easy and that suffering may be the ultimate path to a better future.[105] In Teilhard's system, the suffering of the Cosmic Christ is the suffering of the unthought, and as we have seen, the unthought, so unpleasant to face and so laborious to overcome, is the future evolution of the human. Moreover, Teilhard's opaque, challenging, and poetic writing, while not always easily legible as science, is interested in provoking a wide-open field of Utopian experimentation. Teilhard's writing serves as a provocation to action, as a proleptic materialization of Omega, whereas something like the hedonism of Pearce or the managerial futurism of Huxley's revision of Teilhard merely demands an awed sense of expectation. Teilhard writes a call to noögenesis and performs a model for how such a crowd-sourced global experiment in evolutionary overcoming might be realized.

This experimental method is, I argue, the most important rhetorical pattern Teilhard inaugurates for transhuman thought. His account of biological *and* spiritual evolution, on an equal footing, coalesces a number of precursor elements into a recognizable transhumanism. Despite this, due primarily to a perceived mysticism, what would become contemporary transhumanism almost immediately expels many of the elements of Teilhard's thought, especially the importance of community and suffering to evolutionary change. Despite these ejections, we can see that these elements importantly focus attention on the experimental quality of transhumanism and the associated rhetoric of evolutionary futurism. Teilhard's conjunction of spirit and bios in an attempted scientific account of the spirit further mirrors the growing interest—in the form of, for instance, works by the Dalai Lama and Dorion Sagan—in what might be called a science of the spirit. Despite a growing interest in the shape of his thought, Teilhard's rejection by scientists *and* transhumanists, as Nick Bostrom points out in his transhuman history, does not nor should not diminish his potential contributions to an evolutionary futurist Utopian project.

4 TRANSHUMAN AESTHETICS

The New, the Lived, and the Cute

In the previous three chapters, we have seen how evolutionary futurist themes have an extensive history of critique and construction within aesthetic texts (such as film, genre fiction, and poetry). This engagement with evolutionary futurism in art occurs, however, in spite of the often inartistic aims of contemporary transhumanism. Beyond the allergy (discussed in chapter 2) to being seen as merely science fiction taken too seriously, the ethos of transhumanism is one of science, logic, and rigor. Often this focus on the hard sciences appears antithetical to creative, artistic endeavor. However, in this chapter I argue that there are a number of vectors for exploring evolutionary futurism as a Utopian aesthetic practice. I first consider the transhuman art movement organized by Natasha Vita-More, which argues that evolutionary futurism is a problem that also can be approached through the creativity of the artist. I then move to what I call the "transhuman inaesthetic," present in the work of transhumanists such as Raymond Kurzweil and Zoltan Istvan, showing that, despite Vita-More's efforts, contemporary transhumanism often appears inartistic because of an unwillingness to admit the central role of the sublime in pushing creative evolution. Finally, the chapter considers the New Aesthetic (NA) and online meme culture as two examples of an evolutionary futurist aesthetic practice that point the way toward a creative practice of overcoming the human.

Transhuman Aesthetics

Contemporary transhumanism often constructs an ethos of rational scientism in approaching some seemingly far-out ideas. As Nick Bostrom and David Pearce articulate in their "Transhumanist Declaration," a document associated with the foundation of the World Transhumanist Association in 2003, one major role of transhuman philosophy is "to create forums where people can rationally debate what needs to be done, and a social order where responsible decisions can be implemented."[1] This idea of a rational approach to a reasonable future (though one filled with techno-enhanced immortal humans) represents a key component of contemporary transhumanism's ethos. However, this association has not always been the case. The early formulation of the movement (comprising people like Max More and FM-2030) grows out of the psychedelic and cybernetic futurism associated with Timothy Leary in the late 1970s and early 1980s. Thus, there is a serious core of weird futurism and psychedelic speculation in the founding of transhumanism. The transhuman art movement that Natasha Vita-More pioneered in the early 1980s is one of the best examples of this weirder version of transhumanism. Vita-More, in her two transhuman art manifestos (1983 and revised in 2003), argues that artistic creativity, more than logic, may serve as a core speculative method for manufacturing an evolutionary future. Further, in the work of prominent transhuman architects Shusaku Arakawa and Madeline Gins, we can see an example of how this creative force participates in the creation of a transhuman future.

In 1983, Natasha Vita-More first published the "Transhuman Manifesto," revised in 2003 as the "Transhuman Arts Statement." Both works document an important, emergent engagement with evolutionary futurism as a rhetoric of the future that requires an intense aesthetic component. Vita-More's original manifesto perhaps best hints at the reason such a document was needed at that time. While much of the document, written in poetic free verse, stands as a catalog of evolutionary futurist tropes, the final line suggests that transhuman art, in concert with transhuman science and technology, moves "Toward a more humane transhumanity."[2] As Vita-More suggests in the original manifesto, "Let us choose to be transhumanist not only in our bodies, but also in our values," suggesting a continued focus on technological quick fixes and solutionism stands in opposition to the kind of world that transhumanism (at least, the transhumanism of figures like Vita-More) is trying to build. The command to imagine a transhumanism of values is important for connecting transhuman aesthetics, as Vita-More imagines them, to the broader version of evolutionary futurism I have been tracing in previous chapters. As Vita-More demonstrates, transhumanism needs art because art can teach people how to live as transhumans.

This concept of a transhumanism of values is intensified in the 2003 revision of the manifesto. This later version borrows several of the key lines from the earlier piece but expands and clarifies a number of earlier positions. Of most direct relevance to the question of artistic contribution to the contemporary transhuman program, Vita-More revises the original line "We are designing the technologies to enhance our senses and understanding" into the couplet of "Emotions are integral to our senses and understanding. / We are designing the technologies to enhance our senses and understanding."[3] This revision mirrors the general change in content between the two versions. The 1983 document primarily elucidates the overall goals of transhumanism: life extension, hyperintelligence, and the need for new philosophies of the self for managing these technological changes. In the revision, Vita-More's use of "emotion" marks a rather shocking departure from the tone of the 1983 manifesto. The ethos of the original document imagines art as another vector (along with science and philosophy) for advancing the cause of transhumanism. To wit, the original manifesto did not even contain the word "art" in its title, which changed in 2003. The later manifesto finds Vita-More articulating that art is not just one force among many but that it contributes specific factors, namely aesthetic and emotional ones, to the overall project of transhumanism. Vita-More, in the second version, explicitly argues for the usefulness of art to transhumanism, something that is tacitly assumed in the 1983 document. In other words, we can infer a history in which topics of art and emotion are excluded or ignored over the twenty years that divide the two manifestos; an absence that Vita-More feels called to address in revising the statement regarding transhuman art.

To better see the role of these concerns in practice, I turn to a specific and well-developed example of the kind of transhuman art Vita-More imagines in the manifestos. In the architectural, artistic project that has absorbed much of the creative lives of Arakawa and Gins, we can see in action a version of the aesthetic, emotional transhumanism for which Vita-More agitates. Arakawa and Gins were (he died in 2005; she is still living) a pair of conceptual artists whose quest to actualize a sense of reversible destiny in humanity led them to an artistic practice that increasingly blurred the boundaries between experimental art and hypothetical architecture. In this project, they sought to design environments that would train their occupants how not to die, one of the mantras that repeat throughout their work. In discussing their practice in the concept of a transhuman aesthetics, we can see how the ideas Vita-More outlines have already shaped aspects of artistic practice during the twentieth century. What makes Arakawa and Gins so especially important for exemplifying this is how easily their aesthetic investigations into the human condition begin to resemble time-and-motion studies of the human body during the intensification of manufacturing capitalism at the dawn of the twentieth century. However, where Frederick Winslow Taylor sought techniques for exploring how the human body could be made machinic, the artistic architecture of reversible destiny seeks to instantiate the scientific creation of a postdeath humanity through aesthetic production.

Beginning with their first collaborative project, *Mechanism for Meaning* (1967–77), Arakawa and Gins have sought to better answer the question, "Who or what are we as this species?" The best answer they can offer, found in their later manifesto *The Architectural Body,* is that we are "Puzzle creatures to ourselves, . . . visitations of inexplicability."[4] Their work is suffused with the transhuman trope of faulty biology. They specifically argue that our bodies and our entire culture enforce the notion that death is the inevitable byproduct of living, as do many contemporary transhumanists. For Arakawa and Gins specifically, though, their take on this fault is tied more to notions of destiny and bound up with philosophies of phenomenology. For instance, in *Reversible Destiny* (the catalog of a Guggenheim retrospective of their work), they speculate that "we may be bound to an apparently intractable so-called human destiny simply because we have been unable to gather enough information on our own behalf and to coordinate it properly."[5] They conclude that death may merely be "an inexorably abominable condition" that we are "in the throes of" because we do not fully understand if this end of life is in fact the only destiny awaiting us.[6] This argument—that humans maybe should not so easily and willingly accept death—is not unlike Aubrey de Grey's assertion that biological evolution may be something we can "hack" out of our bodies. However, Arakawa and Gins deviate from this obsession with hacking the body in a way that aligns with Vita-More's claims about the role of the aesthetic in producing a transhuman future.

For Arakawa and Gins, unlike a majority of other transhumanists, the problems of human mortality lie not just in our faulty biology but also in an architectural environment that does not do enough to train us not to die. Specifically, they argue that this malfunctioning architecture lies in our inability to grasp the role that the patterns of collisions, what they call landing sites, play in shaping and conditioning our understanding of ourselves as mortal beings. Similar to other transhuman arguments about a postdeath future of the body, Arakawa and Gins suggest that the human crisis of death emerges from our lack of understanding ourselves and our need for a more thorough investigation of our own existence: "at best, we move in a morass of inconclusive investigations and fragmentary pursuits; at worst, it is assumed that our species will always remain a mystery to itself."[7] This position is one taken up throughout evolutionary futurist rhetoric: if we only had more complete knowledge of our bodies and our minds, we could better begin to hack them into something else. However, Arakawa and Gins deviate from this position in a shocking and shockingly aesthetic manner. Where many contemporary transhumanists advocate hacking the physical body to extend and enhance life, the practice that emerges from Arakawa and Gins's early experiments into meaning suggests shifts in the basic definition of the body itself. The main purpose of their manifesto of aesthetic transhumanism, *Architectural Body,* is unpacking an understanding of the human body as a seamless component of the world itself, a Möbius strip between self and world that mirrors the model for evolutionary futurism found in P. D. Ouspensky. As they declare in *Architectural Body,* "we believe that people closely and complexly allied with their architectural surrounds can succeed in outliving their (seemingly inevitable) death sentences!"[8] They want to argue in their work and in their theories that our built environments, especially the ones in which we live, are part of our bodies and should be considered, definitionally, as such.

Arakawa and Gins: "architecture is the greatest tool available to our species, both for figuring itself out and for constructing itself differently."[9] Throughout their work preceding *Architectural Body,* they refer to this process as one of "reversible destiny," the provocation of "what if it turned out that to be mortal was not an essential condition of our species?"[10] However, as their manifesto makes clear, the condition for reversing this destiny is the built environment. Their method for doing so is to "have the architectural surroundings themselves . . . pose questions directly to the body."[11] This is done in order to manipulate our thought projections of ourselves as on a one-way declination toward mortality. They write: "sentience assembles its swerving suite of cognizing stances depending on how the body disports itself."[12] As they go on to clarify, our lives are an ever unfolding of our attempt to "person" our world (in that we attempt to understand our world through our human needs it addresses), and that "the momentum an organism is able to gain on being a person, or rather, on behaving as one . . . depends directly on how it positions its body."[13]

Part of the architectural research that structures the aesthetic transhuman project of reversible destiny took the form of mapping what Arakawa and Gins call "landing sites." As outlined in the material in the exhibition catalog for their Guggenheim retrospective, *Reversible Destiny,* landing sites are the locations perception picks out for collisions or potential collisions as an organism moves through space. Arakawa and Gins identify three kinds of these landing sites: "some quality of a here or there (perceptual landing sites) . . . between areas of perceptual change . . . a general filling of the gaps (imaging landing sites) and an intimation of position (architectural landing sites)."[14] As they outline in a letter to Jean-François Lyotard, "every bodily motion within an architectural surround elicits a particular constellation of configurations. Changing one or two aspects of an architectural surround—pitch of terrain or general orientation—has the effect of drastically altering a few of a constellation's configurations."[15] Thus, they suggest that in "responding to familiar surroundings . . . the viewer rests comfortably . . . were he confronted with a less straightforward surface, for example, one in which the terrain made him wonder whether he would cross the room, his pose would probably be a more apprehensive one."[16] Terrain that can challenge our expected configuration of landing sites can, as they explain to Lyotard, change the bodily manner in which our tumblings through space produce the concepts that undergird our everyday philosophies of how the world, and our bodies, work.

Thus, the kind of spaces Arakawa and Gins envision as teaching reversible destiny, whether the copious theoretical spaces they designed or the few projects they constructed (including, famously, Site of Reversible Destiny in Yoro, Japan, and their own home, the Bioscleave house, in East Hampton, New York), tend to be multiply variegated spaces with undulating walls that cohere into directed flocks, like swimming fish, while also using tricks of perspective to challenge our assumptions of how spaces should be, including a number of seeming grassy meadows in the Site of Reversible Destiny that are actually optical illusions, with the grass growing on deliberately curved and steeply angled surfaces, or the famous folly in the same site that is a traditional Japanese meditation garden wrapped 360 degrees around the inside of a giant concrete tube. The whimsy that structures these spaces belies the potential brutality of Arakawa and Gins's vision, a vision shared with the Dada practice that partly inspired Arakawa's work. As they make clear in discussing their concept of the landing site, the body in a space configured to teach reversible destiny is "apprehensive."

Arakawa and Gins assert that "No one should consider herself a finished product or a non-puzzle; everyone should live as a self-marmot (self-guinea pig)."[17] To live is to be challenged to live. Describing a hypothetical site visit with clients who might be living in one of their experimental houses, Arakawa and Gins describe the space, which looks like "a low pile of material that covers a fairly vast area" ("this heap?" asks Robert, one of the clients), as made of an ultralight material "developed by NASA" that enables the house's residents to reconfigure the space as needed.[18] As Gins suggests in the site visit's dialog, "the material expands and contracts" in sync with its occupants' breathing; the house becomes an extension of the body, which is key to the concept of architectural body they are expounding. As they explain to the skeptical clients,

> ROBERT: You mean people can actually live here?
> GINS: Of course. I was hoping you would want to.
> ARAKAWA: Live here and do daily research.
> ROBERT: Research into what?
> ARAKAWA: Into what goes into being a person. This place can help you do that.[19]

This view of architecture as creating spaces for mutual coexperience (mapping the house as part of the human body and simultaneously mapping the landing sites this conjunction produces) Arakawa and Gins label "procedural architecture." In their conceptual universe, such an approach to architecture, heretofore untried, can create an escape velocity from the human script that ends in death. However, this process of emergence is seemingly a never-ending process of experimentation, a constant challenge one has to solve in one's day-to-day activities. In this way, Arakawa and Gins arrive at a kind of practice of personhood—an organism "persons the world" as they describe it in *Architectural Body*—that seems somewhere between Henry David Thoreau's desire to live deliberately and Michel Foucault's idea of aesthetic self-fashioning, all routed through the rigorous body control of Pilates.[20]

In any case, their call to live in spaces "tactically posed" to construct "a precise tentativeness" in which one is constantly having to negotiate and renegotiate the architectural body one inhabits is pretty far from the usual *topoi* of the enhanced, technological body to be found in contemporary transhumanism.[21] Through use of and an openness to aesthetic practices of the self and the world, Arakawa and Gins embody Vita-More's call for an aesthetic transhumanism. By putatively abandoning the fetish of scientific rigor, the architectural body and procedural architecture stand as challenges to the notion of human complacency in the face of a seemingly inevitable death, but without the technological determinism and passive faith in emergent technology that one often finds in other, more digitally bent transhuman thinkers. Moreover, Arakawa and Gins, through the strange and variegated tactical spaces of reversible destiny they have built and imagined, remind us that to live forever is fundamentally to live different. Any possible evolutionary future will inevitably be completely different from our present configuration. By radically reimagining space and what it does in our lives, Arakawa and Gins practice a recognizably transhuman notion of overcoming and mastery, capture the profound possibilities of a fully realized evolutionary futurism, and, most important, show the importance of aesthetics in this evolutionary future to come.

Transhumanist Inaesthetics

Despite Vita-More's efforts and work by artists such as Arakawa and Gins, transhumanism has an aesthetic problem. As mentioned in chapter 3, the conceptual figure of evolution's managing director, first established by Julian Huxley, is a major trope in contemporary transhuman thought, appearing prominently in works by Nick Bostrom and Raymond Kurzweil. This figure of the middle manager has all the aesthetic appeal of corporate art: bland, safe, and exceedingly beige. As we saw in the introduction, the primary rhetorical appeal of contemporary transhumanism is logos driven, rather than grounded in the emotional appeals of pathos. A series of logical accounts of scientific wonders that will come to pass is certainly exciting, but it lacks the emotional attachment to the future made possible by the aesthetic. Moreover, as we have seen throughout this volume, the primary effect of logos-driven transhuman argumentation is that of waiting for the emergence of the future, rather than actively desiring its construction.

This logos is most troublingly on display in "Transhumanist Art Will Help Guide People to Becoming Masterpieces," an opinion piece posted by Zoltan Istvan—the Transhumanist Party's candidate for president of the United States in 2016—on *The Huffington Post*'s blog.[22] Istvan's overview of transhuman art touches on artists associated with Vita-More but dedicates most of its efforts to work that helps popularize the arguments and themes of transhumanism. For instance, he highlights novels such as Dan Brown's *The Inferno* and the Johnny Depp vehicle *Transcendence* as important works of transhuman art because, despite being openly critical of evolutionary futurism, these texts popularize the existence of the movement, adding to it an appealing layer of aesthetic polish. Where we saw, in the group associated with Vita-More's manifesto, that a more abstract representation of evolutionary futurist *topoi* is possible, Istvan shows a lack of aesthetic imagination when it comes to form, preferring to highlight works in which characters upload their minds into computers, for instance. As Istvan argues, the goal of his understanding for transhuman art is to have "creative works for admiration and improvement of self."[23] This vision of transhuman art runs counter to the work of Arakawa and Gins and is, I argue, emblematic of the attitude that inspired (perhaps necessitated) the 2003 revision of Vita-More's transhuman art manifesto. While we have been exploring the ways artists *are* contributing to transhuman imagination, Istvan primarily sees art as a realist propaganda engine, solely a tool in the marketing of the more serious, rational work being done by the group.

This focus on realist narratives transmitting transhuman content mirrors the logical appeals that make up the core of much of contemporary transhumanism. As we saw in the introduction, Robert Ettinger justified his early form of transhumanism through a logical appeal that "there are vast segments of the world population that will not concede . . . that it is better to live than to die."[24] For Ettinger, being bound to traditions extends beyond just clinging to faulty or suboptimal technology (he includes an anecdote about Eleanor Roosevelt's amazement that sweepers in India use short-handled brooms and have horrible back injuries) to the very idea of choosing to die. In a postdeath world, Ettinger argues, life will be "an open-ended future" of "soft grass, a fresh breeze, fleecy skies, a cool drink, crisp snow, a warm hand, a familiar voice," as discussed more thoroughly in chapter 3.[25] For Ettinger, aesthetics is limited to comfort and easy living, forever. In this model, the appeal for more life is based on a pattern of clichés not unlike those used to market erectile dysfunction medication. This vision of the good life forever is very marketable, though also more than a little tame. In Zoltan Istvan's piece on transhuman art, his proleptic engagement with his audience anticipates their surprise that a movement steeped in science could be amenable to creative output. As he writes, "transhumanist art seems like an oxymoron to some. Is it possible to combine the scientific nature of transhumanism with creative works for admiration and improvement of self?"[26] As we have seen, however, this particular creative vision for the improvement of self is often as aesthetically rich as a Thomas Kincaid painting.

Pushing beyond clichés about philistine scientists and engineers lacking in social graces, there is a philosophical issue that prevents a more full engagement with the aesthetic within transhumanism. This issue emerges through contemporary transhumanism's often contemptuous engagement with the natural and physical worlds that have classically served as the framework for theories of the aesthetic experience. I trace this out by considering the role of rocks and other base matter in a variety of areas, including Edmund Burke's work on the sublime, Raymond Kurzweil's *The Singularity Is Near,* and Deleuze and Guattari's *What Is Philosophy?*. Ultimately, I suggest that transhumanism, as an aesthetic method, must articulate a kind of postnatural aesthetics: one grounded in the manmade and the technological. Following this section, I will explore two contemporary art movements that create this kind of postnatural sublime.

The moment that best articulates what I label a "transhuman inaesthetic"—the reductionist vision of the world that sees biological life (human and otherwise) only in terms of data, logic, and math—occurs in Raymond Kurzweil's *The Singularity Is Near.* Popularizing Vernor Vinge's notion of technological singularity, Kurzweil's bestselling work was the chief vector for the growing visibility of transhuman ideas in the mid-2000s. In this book, Kurzweil develops an extensively detailed account of how humans may evolve beyond their current confines through a technological coevolution with computers. In addition to articulating logical theories that purport to scientifically prove the inevitability of Singularity—including ideas such as "The Law of Accelerating Returns," which encapsulates his belief that technology is and always will be offering progressively more wonders—Kurzweil offers a variety of philosophical arguments intended to evoke and extend the work done by professional philosophers such as Max More, Nick Bostrom, and David Pearce in articulating transhumanism as a contemporary philosophical endeavor. Kurzweil's postnatural vision is best understood through the frequent discussions of rocks in *The Singularity Is Near.* One key hypothetical technology that will help accelerate the intelligence explosion en route to the Singularity is the ability to convert matter, at a molecular level, to material capable of computation. This theoretical material, called "computronium," emerges from work done in the 1980s on cellular automata by

Tommaso Toffoli and Norman Margolus.[27] For Toffoli and Margolus, the promise of computronium is the promise of infinitely fine-grained simulation of reality, arguing that, "in programmable matter, the same cubic meter of machinery can become a wind tunnel at one moment, a polymer soup at the next; it can model a sea of fermions, a genetic pool, or an epidemiology experiment at the flick of a console key."[28] This concept of programmable matter was seized on by a number of Singularitarians as a means of actualizing an intelligence explosion, especially through the conversion of "dumb" matter to this "smart" matter. For instance, in Charles Stross and Cory Doctorow's linked set of science fiction short stories, *The Rapture of the Nerds,* the extant matter of the solar system has been converted into a giant solar-powered sphere, encircling the sun, that runs an incredibly sophisticated array of human–AI hybrid beings in a post-Singularity reality.

Kurzweil makes extensive use of the Utopian potential of converting all matter in the universe into computronium in outlining his Singularitarian platform. In doing so, Kurzweil asks his readers to consider the uselessness of an average, 2.2-pound rock:

> To appreciate the feasibility of computing with no energy and no heat, consider the computation that takes place in an ordinary rock. Although it may appear that nothing much is going on inside a rock, the approximately 1025 (ten trillion trillion) atoms in a kilogram of matter are actually extremely active. Despite the apparent solidity of the object, the atoms are all in motion, sharing electrons back and forth, changing particle spins, and generating rapidly moving electromagnetic fields. All of this activity represents computation, even if not very meaningfully organized.[29]

He clarifies the rock's prodigious though underappreciated activity by suggesting that,

> despite all this activity at the atomic level, the rock is not performing any useful work aside from perhaps acting as a paperweight or a decoration. The reason for this is that the structure of the atoms in the rock is, effectively, random. If, on the other hand, we organize the particles in a more purposeful manner, we could also have a cool, zero-energy-consuming computer with a memory of about a thousand trillion trillion bits and a processing capacity of 1042 operations per second.[30]

Kurzweil's point here is one that is rather common in Singularity arguments: "when you code activity as computation, everything looks like a computer." For Kurzweil, the question of use is always in terms only of computation and only of use for humans. He considers the rock "to assess just how far biological evolution has been able to go from systems with essentially no intelligence (that is, an ordinary rock, which performs no *useful* computation) to the ultimate ability of matter to perform purposeful computation."[31]

There are a number of problems associated with Kurzweil's rhetoric in this passage. He evinces a reductionist logic in the manner in which he substitutes concepts. In calculating "the computation that takes place in an ordinary rock," "activity" ("sharing electrons back and forth, changing particle spins, and generating rapidly moving electromagnetic fields") becomes "computation." Further, "use" becomes "use for humans," as "the rock is not performing any useful work aside from acting as a paperweight or a decoration." This slippage, this rhetorical substitution of "use" for "use for humans" appears remarkably similar to Martin Heidegger's accusation that technology reduces matter to "standing reserve." Heidegger says that within the rationality of technoscience, "everywhere everything is ordered to stand by, to be immediately at hand, indeed to stand there just so that it may be on call for a further ordering. Whatever is ordered about in this way has its own standing. We call it the standing-reserve."[32] Kurzweil's program of Singularitarianism is a kind of hypertrophied logic of standing reserve: the material world converted into computer.

Rocks, for Kurzweil, are the perfect example of an entirely useless matter: the only thing they do for us is to serve as "a paperweight or a decoration." Their uselessness is a uselessness *to* Raymond Kurzweil (who presumably does not live in a house made of stone or who has never sat on a rock during a hike), specifically. Rocks are not fast, and they are not intelligent: it is our job to reorder their atoms to do computation for us. Kurzweil does show that there are actually two uses a rock could be put to: (1) we could write on its surface and (2) "by dropping the stone from a particular height, we can compute the amount of time it takes to drop an object from that height."[33] For Kurzweil, not only is matter not performing ordered computation useless, any work that is not computation is also useless, as the only two things a rock can perform are rudimentary physics experiments (calculation) and writing (storage). Storage and calculation are also, of course, the two operations performed by a universal Turing machine, of which digital computers are a subset. Presumably humans would never need to use a rock, or anything, for something other than those two tasks. As I alluded to previously, this simple formula ignores the fact that many people eat flour ground using stones, live in houses made from stones, or use tools made from stones. Like a man with a hammer who thinks everything is a nail, part of Kurzweil's appeal lies with the fact that he constructs a model of the world that views it as a big, badly programmed computer just waiting to be debugged.

Kurzweil's transhuman inaesthetic, then, is a philosophy without the sublime. Edmund Burke, in *A Philosophical Enquiry into the Origin of Our Ideas of the Sublime and Beautiful,* defines the sublime as "the strongest emotion which the mind is capable of feeling" and remarks that these feelings of intense beauty and fear are provoked by "whatever is fitted in any sort to excite the ideas of pain and danger, that is to say, whatever is in any sort terrible, or is conversant about terrible objects, or operates in a manner analogous to terror."[34] The philosophy of sublimity is long (to put it lightly), stretching from Longinus through Burke and Kant's commentary on Burke into the present. While I will not rehearse that history, I will briefly discuss Burke's work because of its importance to Romantic poetry and that poetic movement's signature appreciation of the feelings of terror and beauty provoked by the natural world. Romanticism is often cited as inventing the modern idea of the tortured artist–genius, a kind of mad creativity that inheres in Vita-More's version of transhuman aesthetic but is a worldview almost completely alien to Istvan and Kurzweil. Moreover, I look at Burke because in a number of places in *The Origins of the Sublime and Beautiful* he explicitly discusses rocks as sources of the sublime. Thus, I want to use Burke's account of "the strongest emotions which the mind is capable of feeling" as a rebuttal to Kurzweil's utilitarian discussion of rocks.

In cataloging the aspects of the natural and built worlds to induce feelings of the sublime, Burke suggests in his section on vastness,

> Of these the length strikes least; an hundred yards of even ground will never work such an effect as a tower an hundred yards high, or a rock or mountain of that altitude. I am apt to imagine likewise, that height is less grand than depth; and that we are more struck at looking down from a precipice, than looking up at an object of equal height; but of that I am not very positive. A perpendicular has more force in forming the sublime, than an inclined plane; and the effects of a rugged and broken surface seem stronger than where it is smooth and polished.[35]

As he documents, the experience of height produced by mountains (and in the present by built objects such as skyscrapers) suggest a scalar relationship in which our small and limited human form is found wanting. Citing Psalms 114:8 ("Tremble, thou earth! at the presence of the Lord; at the presence of God of Jacob; which turned the rock into standing water, the flint into a fountain of waters!"), Burke explains that such passages ("endless to enumerate") "establish the general sentiment of mankind, concerning the inseparable union of a sacred and reverential awe, with our ideas of the Divinity."[36] This notion of awe also structures the encounters with rocks and stones in the natural world. Discussing Stonehenge, Burke suggests that "those huge rude masses of stone, set on end, and piled each on other, turn the mind on the immense force necessary for such a work," again reminding humans of their limits.[37]

The absence of sublimity in Kurzweil hinges on this experience of a limit. For Burke, and others who have written in the long history of the sublime, the strongest emotions provoked by experiences of vastness confront us as individuals marked by our limitations: by our individuality and by the brevity of our lives, the smallness of bodies, and the fragility of life itself. Contemporary transhumanism confronts these same factors but, as we see in Kurzweil's unwillingness to engage with thinking of rocks in terms of the sublime as Burke does, must do so in a way that ignores human frailty. Transhumanism is a version of evolutionary futurism marked by an overwhelming optimism and a belief in the limitless capacity for human transformation. While it seems that a version of transhumanism that utilized the sublime as a vector toward an evolutionary future would be a hugely successful rhetorical strategy for actualizing a transhuman future through the use of our experience of awe as a source of desire for transcendence, such a move has not occurred. Contemporary transhumanism, without limits, is a vision of the universe without the sublime.

However, an approach to evolutionary futurism that accounts for the sublime can be found in the works of Gilles Deleuze and Félix Guattari. In their final work, *What Is Philosophy?* we can find an alternate approach to human overcoming and evolutionary futurism that has a very different account of "useless" matter. For as much energy as Kurzweil expends on mapping the computational potentials of uploaded rocks, Deleuze and Guattari's account of consciousness in "From Chaos to the Brain" also hinges on the computational potential of rocks, though in a very different way. They write, in a vein similar to Arakawa and Gins's discussion of the human "personing" their world, of how, as a unit of selection, species extend through the world long causal chains (for plants: "light, carbon, and the salts") that result in "sensation in itself" and come to constitute something that looks a lot like mentality.[38] From this, however, they conclude,

> Of course, plants and rocks do not possess a nervous system. But, if nerve connections and cerebral integrations presuppose a brain-force as faculty of feeling coexistent with the tissues, it is reasonable to suppose also a faculty of feeling that coexists with embryonic tissues and that appears in the Species as a collective brain; or with the vegetal tissues in the "small species." Chemical affinities and physical causalities themselves refer to primary forces capable of preserving their long chains by contracting their elements and by making them resonate: no causality is intelligible without this subjective instance. Not every organism has a brain, and not all life is organic, but everywhere there are forces that constitute microbrains, or an inorganic life of things.[39]

This paragraph represents a counterpoint to Kurzweil's account of material and computation. For Kurzweil, rocks can only attain any kind of usefulness for consciousness through their conversion by humans into computronium. For Deleuze and Guattari, the entirety of being constitutes a seething "inorganic life of things" that pulsates with not only a life but a kind of universal consciousness, constituted by what they call "microbrains." These microbrains represent the culmination of a series of "micro-" concepts in Deleuze and Guattari's collective oeuvre, including "microstructures," "micropolitics," "microfascisms," and "microperception." For Deleuze and Guattari, as Brian Massumi explains in an interview with Joel McKim, the micro, encapsulated by the master term "microperception," is "something that is felt without registering consciously. It registers only in its effects."[40] A microbrain, then, would be a sense of consciousness that does not, in and of itself, actively register but churns below the level of human awareness, though theoretically it could be inferred by an investigation into apparently inexplicable effects.

Deleuze and Guattari's account of intelligence in the above passage suggests an understanding of life that hinges on a universal distribution of Brain as a globally binding force. For Deleuze and Guattari, the brain is not reducible to an individual brain (the goo inside our skulls): "it is the brain that thinks not man—the latter only being cerebral crystallization. We will speak of the brain as Cézanne spoke of the landscape: man absent from, but completely within the brain."[41] The idea that humans are merely "cerebral crystallizations" suggests this universal aspect. This idea of cerebral crystallization provides a fuller portrait of the "microbrains" that constitute the "inorganic life of things" in the selection above. For Deleuze and Guattari, while humans are cerebral crystallizations, so are rocks, plants, bacteria, tables, coffee beans, and so on. The "inorganic life of things" structures an ontology in which Thought precedes Matter. We can verify this proposition by considering their description of a universal Brain that "is not a brain behind the brain, but, first of all, a state of survey without distance, at ground level, a self-survey that no chasm, fold, or hiatus escapes."[42] For Deleuze and Guattari, Brain is everything, an unending, ongoing unfolding of reality that is continually computing itself.

For Deleuze and Guattari, the intelligence of rocks is thus a reflection of Brain, the "single plan of composition bearing all varieties of the universe."[43] In this discussion of rocks, plants, and the Chaos of creation, Deleuze and Guattari are applying a previously unseen sense of urgency to their more well-known arguments about the univocity of Being. More important, though, this argument for an inorganic life of things composed of microbrains suggests a powerful counterpoint to Kurzweil's claims about rocks that we saw above. For Deleuze and Guattari, matter itself is already embroiled in computation, because, if for no other reason, this inorganic life of things represents a provocation to thought. Deleuze and Guattari situate their understanding of Brain as a counterforce to Chaos, the flux of an unruly Being. *What Is Philosophy?* is taken up with analyzing the three planes of concept formation—art, science, and philosophy—and in "From Chaos to Brain," they suggest that these three planes, which meet but do not unite in Brain, are allies in the struggle against Chaos: "people," they write, "are constantly putting up an umbrella that shelters them and on the underside of which they draw a firmament and write their conventions and opinions. But poets, artists, make a slit in the umbrella, they tear open the firmament itself, to let in a bit of free and windy chaos."[44]

This "free and windy chaos" is what is lacking in Kurzweil's account of matter. As we have seen in this section, Kurzweil's account of "dumb" matter simplifies thought to mere computation, itself a simulation of thought, in a metaphoric forgetting of the fact that humans created computers as crystallizations of certain operations run by our more complex brains. Deleuze and Guattari's account of Chaos and Brain, as macroformations exceeding individual humans or individual environments, suggests a complex boldness that characterizes thought, outside Kurzweil's clinical and antiseptic model. Moreover, by accounting for the "inorganic life of things," Deleuze and Guattari remind us of the inherent situatedness of human thought: an intellectual and evolutionary legacy developed specifically in response to the provocations of the dumb matter Kurzweil finds so little use for. In evolutionary biology after Darwin, thought emerges as a response to the need to solve such specific, environmental problems and cannot so easily be divorced from this lived environment.

Even in something as lowly as a stone we can find the kind of confrontation with chaos that sparks the creation of new concepts. In establishing a dialogue between Kurzweil and Deleuze and Guattari on the topic of the sublime, we can begin to imagine the role the sublime might play in creating an evolutionary futurist aesthetic. Where Vita-More's work imagines the future as a creative problem that must be solved through the artistic impulse, an evolutionary futurist aesthetic driven by the sublime would instead magnify the bigness and the oddness of our own era of rapid technological change. Such an aesthetic would need to magnify the scope of our world to highlight the evolutionary imperatives shaping our experience of that world. It would magnify our sense of ourselves within our mutation. In what follows, I consider two art movements, one from high culture and the other not, that offer us glimpses of an evolutionary futurist sublime. In both the New Aesthetic in high art and online meme culture, we see attempts to map the future in aesthetic terms but, moreover, use the tools we find to do this mapping as sources of the intense emotions that make up the experience of the sublime. Whether hanging geometric satellite images in art galleries or circulating talking cats online, both of these aesthetic formations map an evolutionary futurist aesthetic that, I argue, better articulates an art of an evolutionary future than any of the attempts by contemporary transhumanism to manifest as an art movement.

At Home with the Future: The New Aesthetic as Everyday Transhumanism

The New Aesthetic, compared with the rest of the cultural phenomena investigated here, only barely exists. Like any truly interesting thing happening in the world today, it shimmers as an interlinked cloud of blog postings, tweets, and posts to various Tumblrs. It was also the subject of a panel at South by Southwest, that official gatekeeper of what counts as "cool" in our present reality. On April 2, 2012, the New Aesthetic was formally announced when, following this SXSW panel, Bruce Sterling posted to his blog for *Wired* magazine, *Beyond the Beyond,* an article titled "An

Essay on the New Aesthetic." Within four days, members of the Creators Project, an art collective whose members' work is often lumped under the umbrella of this emerging movement, had posted an anthology of essays responding to Sterling's piece. After that, the New Aesthetic began to have more of an existence, but it was still very hazy around the edges, as it is not characterized by the organizational efforts that usually shape art movements.

The movement takes its name from a Tumblr maintained by James Bridle, a graphic designer and artist from London. Bridle, along with several friends, began cataloging blog images and videos that, to them, document a genuinely new aesthetic sensibility developing out of the increasingly mediated nature of our existence. The site catalogs images that relate to things like military drones, computer vision, image-processing algorithms, and, especially, Google Maps. A recent post, for instance, detailed two moments when the BBC used screen captures from video games—from *Assassin's Creed* to represent Damascus and the logo of the United Nations' Space Counsel from *Halo*—as over-the-shoulder images during its nightly newscast. Generally, this curio cabinet of Internet delights documents the idea that there is something very odd at work in our aesthetic sensibilities. As we shall see in this section, this exploratory and curatorial approach isolates the New Aesthetic as an exploratory project of cognitive mapping for an emerging transhumanity. Specifically, this project maps the ways we are increasingly domesticating our relationships with machines.

In a 2011 post by which Bridle's blog inaugurated the New Aesthetic Tumblr, the artist lays out some of the basic hypotheses of the New Aesthetic. In this post, he connects these images and video clips to a changing sense of futurity itself:

> For a while now, I've been collecting images and things that seem to approach a new aesthetic of the future, which sounds more portentous than I mean. What I mean is that we've got frustrated with the NASA extropianism space-future, the failure of jetpacks, and we need to see the technologies we actually have with a new wonder.[45]

The kind of future projected equally by NASA's rhetoric during the mid-twentieth century and by cultural institutions like *Star Trek* is, as Bridle suggests, frustrating in its elusiveness, its tendency to recede indefinitely. We *still* do not have the jetpacks or cities on clouds that were promised us by pulp science fiction in the 1920s, yet despite the exhaustion of the vision of the future embodied by the jetpack, Bridle suggests that we are still living in *a* future, even if it is not the future projected by the American technopositive science fiction written before 1980.

Unlike the cynical rejection of this future projected by cyberpunk authors such as William Gibson, especially in "The Gernsback Continuum," Bridle calls for a "new aesthetic of the future" that is also a call to reactivate a dormant sense of wonder about technology. As an example of the rhetorical shift attempted by New Aesthetic practice, consider the following joke from comedian Louis C. K. about air travel:

> "I had to sit on the runway for forty minutes." Oh my god, really? What happened then, did you fly through the air like a bird, incredibly? Did you soar into the clouds, impossibly? Did you partake in the miracle of human flight and then land softly on giant tires that you couldn't even conceive how they fucking put air in them? You're sitting in a chair in the sky. You're like a Greek myth right now.[46]

Just as Louis C. K.'s joke works to defamiliarize "banal" contemporary technologies, Bridle's argument in constituting the New Aesthetic is to highlight this newness, the magical nature of our own world. He wants to remind us that we are already living in a science-fictional world, even if, because of our closeness, we do not recognize it as such.

To explore this new world we find ourselves in, Bridle suggests three aesthetic sites for future exploration:

> For so long we've stared up at space in wonder, but with cheap satellite imagery and cameras on kites and RC helicopters, we're looking at the ground with new eyes, to see structures and infrastructures.
>
> Representations of people and of technology begin to break down, to come apart not at the seams, but at the pixels.
>
> The rough, pixelated, low-resolution edges of the screen are becoming in the world.[47]

What is interesting about satellite imagery, glitches, and pixelation is that NA practitioners, in discussing their work, do not have the same detached and ironic relationship to their subject by which, say, Warhol related to soup cans. As Creator's Project member Kyle McDonald explains in his essay on NA, "we borrow the aesthetic of satellite views for our work not because we've been counseled by the machine, but because we have personally judged the results of this functional system as beautiful."[48] It is no longer an ironic celebration of pop junk as art, so much as a legitimate acknowledgment that the militarized gaze of the satellite functions as the new picturesque.

For Sterling, writing in the blog post that prompted McDonald and other members of the Creator's Project to write energetic responses defending NA:

> The New Aesthetic is constructive. Most New Aesthetic icons carry a subtext about getting excited and making something similar. The New Aesthetic doesn't look, act, or feel postmodern. It's not deconstructively analytical of a bourgeois order that's been dead quite a while now. It's built by and for working creatives.[49]

By suggesting that the New Aesthetic is beyond deconstructive gestures, Sterling also wants the New Aesthetic to be representative of a new program beyond the cynical, ironic distance of postmodernism (which is, after all, emphatically over as far as cultural movements go). For Sterling, this is partly due to the creative expression of a new generation:

> It is generational. Most of the people in its network are too young to have been involved in postmodernity. The twentieth century's Modernist Project is like their Greco-Roman antiquity. They want something of their own to happen, to be built, and to be seen on their networks. If that has little or nothing to do with their dusty analog heritage, so much the better for them.[50]

Sterling's blog post was so troubling to his respondents because, while praising their jettisoning of irony and their embrace of creation, he also attempts to write them into a normative generational history of art, in which new, young artists attempt to shock their elders. To that end, he introduces the New Aesthetic to his readers as "a typical avant-garde art movement that has arisen within a modern network society."[51] So, in jettisoning these artists from history, Sterling is also simultaneously even more firmly positioning NA within a supposedly natural and linear progression of avant-garde art as a series of shocks and countershocks. To do this, Sterling ignores the fundamental claim of NA that something basic has shifted in our relationship to art and technology.

It is difficult to situate the New Aesthetic within the history of avant-garde art movements, however, partly because it is so new. Moreover, the people who practice it seem different. As Sterling acknowledges in his blog post, this is a movement made for and by "working creatives," the very kind of knowledge workers who make up the cognitariat under contemporary neoliberalism. Many New Aesthetic projects have as much in common with software development as they do with traditional artistic practices. Significantly, the artists working in this movement or aligned with it spend their days as web developers or other information workers. "Waving at the Machines," Bridle's talk on the New Aesthetic—now regarded as a key moment in shaping the ideology of the movement—was, for example, delivered as the keynote presentation at a web developers' conference.

At the same time that NA does not look like what we are culturally conditioned to think of as avant-garde art, NA does work to change the familiar ways we see and relate to the world. As artist Marius Watz argues in his submission for the Creators Project's collection of responses to Sterling:

> The "New" part is deceptive, however. Most of what NA offers up for examination is not all that new. Technologies like machine vision and geo-location are old by most standards. What is new is their integration into our lives to the point where we are bringing them to bed. Smartphone habituées will think nothing of installing a sleep-tracking app and putting their phone on the mattress, where accelerometers will presumably make sage observations about your quality of sleep. This is the new Aesthetic—human behavior augmented by technology as often as it is disrupted.[52]

Watz here draws out some of the central questions structuring NA. The way humans increasingly trust algorithms and objects to make decisions for them or to answer questions we previously could never have asked represents a basic change in our relationship to technology. The intimacy Watz describes is disturbing to be sure, but it also points out a fundamental change in our understanding of ourselves: we are objects of our own study as much as we are enhanced by these interactions. Watz continues, "The New Aesthetic is a sign saying 'Translation Server Error' rather than 'Post Office'. The New Aesthetic is faces glowing ominously as people walk down the street at night staring at their phones. . . ."[53] The modernist avant-gardes discussed in chapter 1 sought to imagine new relationships between the body and technology and then, as a radical pose, sought to inhabit the seemingly horrifying positions implied by these new relationships (especially in F. T. Marinetti's brand of futurist provocations). However, as Watz draws out in his statement, the radical new technological world that came out the other end of the twentieth century is at once more banal (translation errors taken for translations) and more horrifying (the shuffling zombie of the cellphone user). As Kyle Chayka clarifies in his submission to the Creators Project's responses, "instead of shocking society, New Aesthetic art must respond to a shocked society and turn the changes we're confronting into critical artistic creation."[54] In both cases, Chayka and Watz respond to Sterling by highlighting that what has changed to produce a new aesthetic response is not our technology but our uses of it. Fundamentally, then, NA is a version of Vita-More's transhuman art in practice.

The increasingly intimate relationship with technology documented most directly in Watz's selection, in which users take their mobile phones to bed with them and experience no discomfort in being continually monitored by their devices (all in the name of lifehacking, of course) points toward something truly odd going on. If we were to compare, for instance, the manifesto writings of the New Aesthetic to earlier art movements focusing on technology, specifically the Futurists, I think we would see how truly different this movement is.

As Hal Foster has documented in his exploration of modernism and the machine, "Prosthetic Gods," Wyndham Lewis's Vorticist manifestos and F. T. Marinetti's Futurist manifestos document a relationship with technology in which

> modernist practice and marxist discourse still treated the body and the machine as separate entities, with the first often projected as a natural whole, the second as an autonomous agent. So opposed, the two could only conjoin, ecstatically or torturously, and technology could only be a "magnificent" extension of the body or a "troubled" constriction of it.[55]

As Foster documents in this essay, earlier avant-garde art movements, especially the ones that Sterling conflates with the New Aesthetic, documented a very different relationship between technology and the body than we find operational in the intimate integration of staggering amounts of information into our daily lives. Foster suggests that there are two responses to the modernist vision: a "neoclassical" response that seeks to recuperate previous models of body–technology relations and a "machinic" that glamorizes the violence done to the body by technology. However,

> If the neoclassical reaction proffered the nostalgic balm of an imaginary body that was pellucidly intact, the machinic reaction looked to the very mechanization of the modern body for a new principle of corporal order. At base, however, the first reaction was no more "humanist" than the second: both tended to treat the body as if it were already dead, an uncanny statue in the

> first instance, an uncanny mechanism in the second—that is, *as if the only way for the body to survive in the military-industrial epoch of capitalism was for it to be already dead, in fact deader than dead.*[56]

Foster then suggests that the twin specters of war and industrial labor shape and foreground this deadening of the body in the cultural imaginary: both neoclassical and machinic modernist responses to technological change "are haunted by the specter of the damaged body of the worker-soldier."[57] For Foster, these avant-garde movements sought to imagine new, better humans who could handle the forces of technology that destabilized our humanity. This moment is, of course, the same moment discussed in chapter 1 as one of the origin moments for evolutionary futurism.

This relationship—in which the machine is a monstrous and mangling other—is why I echo several of the artists featured in the Creators Project's response to Sterling in being wary of aligning NA with historic avant-garde movements. Rather, whereas artists such as Wyndham Lewis and F. T. Marinetti, sitting on the other side of the vast sea changes of the twentieth century, wanted to imagine a humanity that could withstand the intensified processes of modernization to come, the New Aesthetic asks questions of how to make sense of these changes that have already happened. We are already through the mangling of the body envisioned by the early twentieth-century avant-gardes that first inspired a series of intensifying responses. We have become the bodies they imagined, though we do not recognize this because our experiences do not match the aesthetic vision of artists like Lewis and Marinetti.

As an example, one of the questions that circulates on the Internet periodically, and that relates to the New Aesthetic's obsession with satellite imagery, is the question of why, when people first encounter Google Maps or Google Earth, the impulse is to not, say, look at London or Tokyo, but to find their houses: "Look, there's our house . . . from space." The idea of using a satellite to tell you what you already know (what your house looks like, where you live) is exceedingly strange and a condition that is uniquely contemporary. The New Aesthetic, conceived as a theoretical approach to making new art with technology, attempts to make sense of this impulse. We are increasingly comfortable with asking technology questions that help us determine our sense of self, but do we understand the answers we receive? What do we know about our role in the world by seeing our house from the perspective of a spy satellite? While, as Chayka suggested, previous avant-garde artists sought to shock people into seeing technology's hold over our minds, the New Aesthetic seeks explanations for these strange and incomprehensible answers we receive from the technologies we use to ask about our world. Like Vita-More in the 2003 revision of the transhuman art manifesto, we need aesthetics to make sense of our actual, existing transhuman lives.

Acknowledging this embrace is central to New Aesthetic practice, but, as Kyle McDonald does in his response to Sterling's essay, we can also acknowledge that "we borrow the aesthetic of satellite views for our work not because we've been counseled by the machine, but because we have personally judged the results of this functional system as beautiful."[58] It is one thing to see this embrace of technology; it is another to find it pleasurable. Pushing into the latter realm is what, I think, marks the emerging New Aesthetic as a possible first wave of a transhuman aesthetic response to the forces of neoliberal control.

As James Bridle discusses in his talk on the New Aesthetic, "Waving at the Machines," this embrace is already beginning to be experienced as the kind of coevolution suggested by evolutionary futurism. In Bridle's vision, however, this coevolution is significantly more banal than the one found in much of contemporary transhumanism. In explaining why the New Aesthetic is everywhere emerging, Bridle speculates about the possible future emerging from cultural productions he and his friends have curated under the NA banner:

> This is waving at the machines. You can foresee a future when in entering a room this is what you'll do, to identify yourselves not just to the people but to the computers and the machines who are watching us too. We'll have entered into this dialogue with them, and we're already doing it like this. We already share our world with these things that are watching us. And it can be creepy and it can be surveillance, or it can be a shared vision.[59]

The phenomenon of waving at the machines places us very far from the kind of violent and revolutionary rhetoric we see in contemporary transhumanists such as Kurzweil and Moravec. In Bridle's vision of a machinic future, technology is increasingly domesticated, somewhere between a bourgeois cyborg and a new kind of digital pet. Of course, at the same time that Bridle's vision suggests a kind of domestic intimacy with our digital devices, in which we wave at them as much as at our loved ones and our pets, the majority of "Waving at the Machines" is spent cataloging the increasingly bizarre effects of this domestication.

Bridle's model of transhumanism is perhaps more accurate, phenomenologically speaking, than the rhetoric of the radical break proposed by Kurzweil's Singularitarian movement. The human mind, after all, is extraordinarily plastic and adaptable. The new normal will look an awful lot like the old normal, until, one day, we realize that everything is suddenly different. Terry Bisson's essay "The Singularity," a refutation of the ideas of Raymond Kurzweil and his Technological Singularity, offers a similar, explicitly transhuman account of this new sense of wonder. Bisson's essay, reflecting on the idea that we are on the verge of a radically new world brought about by an accelerating rate of technological change, ultimately concerns his grandmother and her slow decline as a result of Alzheimer's disease. Reflecting on her life and what she saw, he writes:

> We will die, you and I, in the world we were born into; not so my mother, your grandmother, our great uncle Jim. They were born into a world lit by fire, pulled by horses, or steam at best, and they died in a world knit together electronically, in which no two cities in the world are more than a long day's journey apart.[60]

For people whose lives were rewritten by wondrous changes, the Singularity already happened with the move from the country to the city and the emergence of various global media (telephones, Internet, jet travel). For Bisson, the fact that his mother, as a girl, was lucky to get to the nearest city once a year (on an overnight boat ride) and that, by the end of her life, she could receive calls from her daughter in Thailand, marks the twentieth century out as the true Singularity. He writes of his mother, her memories slowly vanishing to Alzheimer's Disease:

> One night I took her for a drive on the bypass that now encircles her small town: Walmart, McDonald's, 7-11, all ablaze with neon signs. The traffic flowed like a river of light, and Elvis was on the radio, twenty years after his death. The car phone rang; it was my wife, reminding me to pick up ice cream.
>
> My mother sat up suddenly, looked around, delighted, and asked, "What happened here?"
>
> What happened was the Singularity. It happened in an instant of historical time, and it created a world unrecognizable to the little girl who saw it begin.[61]

As Bisson documents, a radical technological shift happened in the change from the village to the city. As Bisson continues, the Singularity "was and is truly wonderful, and it's ours to finish, to refine, to enjoy, and hopefully to learn to control and use. Not ours to create, ours to inherit."[62] This view of a rapid technological explosion in the early twentieth century is verified by historian of science Vaclav Smil, who claims that the 1880s was "the most innovative decade of human history"[63] For Bisson, the accelerating returns of technological change that mark the dawn of the twenty-first century are nothing more than refinement: "Computers get smaller and faster and smarter but they are still just off-loads of ourselves, memory and math. They will never tell us anything that we haven't told them to tell us."[64] What Bisson sees in the radical change caused by the move from the village to the city is a technological singularity that has not, yet, been accompanied by an evolutionary singularity. The project of the human species in the twenty-first century is, as Bisson writes, "to finish, to refine, to enjoy, and hopefully to learn to control and use." Our technology has evolved, now so must we.

Waving at the machines is the moment where we all become like Terry Bisson's mother: looking around in wonder at the world we have made and asking, "What happened here?" The New Aesthetic is obsessed with pixelation, glitches, and satellite imagery. It is obsessed with social media, with big data, with computer vision, with camouflage, privacy, and the extrusion of the Internet into everyday life. These are the pieces of our banal everyday lives, but they are also incredibly strange and the stuff of science-fictional daydreams. While we do not yet appear to have a coherent aesthetic program to make sense of these things, the New Aesthetic suggests one possible response to this new world. Further, as Bridle reminds his audience at the end of "Waving at the Machines," the key to producing a transhuman aesthetic response to the world of big data is to maintain a sense of wonder and awe:

> And my only message is that some of this stuff is completely awesome, and you should always remember that, but also that we should go out there with this willingness and friendliness to engage with technology, to engage with all these technologies while understanding how they shape our behaviors and our feelings and our culture all the time. These things are radically transformative. We are creating a new nature in the world. It's going to be really exciting. Please make it more exciting.[65]

The artistic experimentations of the New Aesthetic most accurately represent the moment we can begin to imagine being beyond the irony, partiality, and social constructionism of the postmodern. The New Aesthetic, rather than continuing to tear at the fabric of philosophical modernism, which, as Bruce Sterling points out in his essay on the New Aesthetic, has "been dead quite a while now," instead signals a desire to build a new aesthetic order to make sense of the world around us.[66]

In the production of this aesthetic response to new technology, the New Aesthetic represents a chance to manufacture an everyday transhuman aesthetic, a worldview in which we can make sense of the need to ask spy satellites about our place in the world. This transhuman aesthetic is, I think, the possible follow-up to postmodernism and the aesthetic regime of our time. As another exploration of this new, emergent aesthetic order, I want to turn from the world of high-art theorizing to the world of low-art production. In exploring the youth culture represented by online meme images, I hope to show another realm in which a transhuman aesthetic is taking shape.

I Can Has Enlightenment? LOLCats and the Global Brain

The New Aesthetic attempts to make aesthetic sense of the increasingly intimate relationships we have with our machines. Additionally, this aesthetic sense helped respond to the conditions of possible global encephalization represented by telecommunications and the rewiring of our cognition. In this section, I further consider the relationship between this natal global brain and aesthetic responses to our present condition. Where the New Aesthetic is very directly tied to the world of gallery art, there is a twin phenomenon occurring online that is not widely recognized as art in the traditional sense. The meme image, a collective pop art project popularized on the vast subcultural Internet forums Something Awful, 4Chan, and Reddit, documents a similar aesthetic response to the evolutionary pressures of telecommunications and textual prosumption; however, if anything, memes, when considered from a transhuman perspective, are perhaps even more experimental and more explicitly transhuman than the New Aesthetic's very domesticated approach to technological emergence. In this section, while I consider memes as a broad phenomenon, I specifically focus my analysis on a meme series called LOLcats. These images—pictures of cats captioned with grammatically incorrect text typeset in the aesthetically horrible font known as Impact[67]—capture an increasingly global, emergent linguistic phenomenon in which we articulate our own relationship to this newly networked world that always seems to just exceed our understanding.

The phenomenon of LOLcats is a form of an "image macro" that has proliferated on the Internet since around 2005. Image macros, according to *Wikipedia,* "are used to emphasize a certain phrase (often an Internet meme) by superimposing it over a related picture."[68] A "meme" is an idea that spreads like a virus and was coined by Richard Dawkins in *The Selfish Gene,* but it comes to mean, specifically, an image or video that, because of its great humor, circulates through culture like a disease. As the *Wikipedia* entry also points out, the term originated on the forum site Something Awful as a funny way to post a prefabricated response to another forum message:

> The name derived from the fact that the "macros" were a short bit of text a user could enter that the forum software would automatically parse and expand into the code for a pre-defined image, relating to the computer science topic of a macro, defined as "a rule or pattern that specifies how a certain input sequence (often a sequence of characters) should be mapped to an output sequence (also often a sequence of characters) according to a defined procedure."[69]

Essentially, an image macro combined a funny statement with an appropriate picture to produce an amusing way to respond through canned text. An example of an early image macro from Something Awful features the picture of a surprised looking owl and the text "O RLY?" (a condensation of "Oh, Really?"). This macro is used as a humorous way of expressing incredulity at an absurd statement made in conversation in a forum discussion thread. On the forum, if someone typed a specific phrase into a post, something like ":orly:," the HTML code for the image would be automatically substituted for the original macro when the message is posted.

LOLcats began as a specific subgenre of image macro on another Internet forum, 4Chan, as part of something called "Caturday," itself a protest against another forum event, called "Furry Friday," in which users post images of anthropomorphized animals engaged in various sexual acts.[70] While "Caturday" began with the general posting of images of cats in cute or funny situations, the practice of captioning the photos soon became commonplace. From these "Caturday" events, the idea of captioning cat images and posting them on the Internet spread and culminated in the creation of a website dedicated to the LOLcat phenomenon called ICANHASCHEEZBURGER. The origin of this title was taken from the caption of one of the first LOLCat images, a picture of the so-called Happy Cat image (taken from a Russian website for the cat food manufacturer of the same name) captioned with the question "I CAN HAS CHEEZBURGER?" The name for this specific genre of image macro is taken from the common Internet practice of typing the phrase "lol" (short for "laugh out loud") to express a moment of intense humor. As these cat images are meant to produce such laughs, the name caught on quickly. As fantasy author Lev Grossman observes, the LOLcat phenomenon is interesting because of

> how little else like it there is online right now. The great, weird Internet meme, which once thundered across the Net in vast herds, is becoming surprisingly scarce, which may be why LOLcats has a distinctly old-school, early 1990s, Usenet feel to it. It's not based on a Saturday Night Live sketch, and nobody's using it to get famous or sell anything. Yet.
>
> We may be witnessing a revolution in user-generated content, but the more mainstream the Web gets, the more it looks like the mainstream: homogenous, opportunistic, and commercial. It's no longer a subculture; it's just the culture.[71]

Grossman positions LOLcats as an anticapitalist, spontaneous expression of human creativity. Extending the culturally avant-garde dimension of LOLcats, a transhuman perspective on the phenomenon suggests an evolutionary role for these images. LOLcats are unique among Internet memes in that the phenomenon seems only to be gaining momentum, even after so many years (especially measured in Internet time). Where early Internet memes, such as "All Your Base Are Belong To Us,"[72] essentially circulated and then vanished, LOLcats suggests the creation of an entire vocabulary, almost a new language for talking about culture. Where earlier memes, such as "All Your Base," certainly solicited user participation, the meme itself only ever consisted of the initial video of fractured, badly translated Japanese. With LOLcats, though, we can speak of an entire collection of memes, even pointing out that LOLcats has spawned several cat-related submemes (or genres), such as the "Invisible *X*" in which freeze-frames of cats in mid-leap appear to be riding bikes or doing other human activities, or the "I'm in ur *X*, *Y*-ing ur *Z*" where cats are photographed in compromising or unusual situations and captioned to suggest they are doing something deliberate or amusingly sinister (such as "IM IN UR FRIDGE EATIN UR FOODZ"). In these cases, Internet users create new images that conform to these submeme forms. All of this suggests that rather than the sharing of a single cultural artifact over and over again, as with the "All Your Base" meme, LOLcats can be thought of as an entire genre of meme that encourages new creations that nonetheless follow specific patterns serving as generic types within the broader generic field.

Extending this idea a little further, we can begin to suggest that LOLcats function as a kind of language, broadly defined, for an emerging global consciousness. I use "language" in the way Roland Barthes uses it in *Mythologies* as "any significant unit of synthesis, whether verbal or visual."[73] So although the texts of the LOLcats genre do not appear to conform to language in the sense of a grammar or a codified syntax, they are nevertheless clearly a specific way of speaking—the stilted speech, the repeated and standardized sentence patterns—and, more important, a way of signifying through image and text. In fact, a team of volunteers recently translated the Bible into LOLcat images.

While one could claim that LOLcats is a heartening phenomenon because this language of a collective human brain is, so undeniably, cute and fuzzy (suggesting that, maybe, we are not as violent as the media leads us to believe), it is also inherently childish, as are many things that happen on the Internet. How can we claim a positive development in a global brain, when that development takes the form of badly misspelled captions under pictures of cute and fuzzy animals?

Quite simply, in fact. One of the aspects of so much debate about global intelligence, especially those influenced by Teilhard, is the idea of the spontaneous emergence of a fully formed, mature global brain. For instance, in Peter Russell's *The Global Brain,* we find an account of the emergence of the "Gaiafield" (Russell's term, coined after rejecting "noösphere" from Teilhard and "Supermind" from Sri Aurobindo):

> As the communication links within humanity increase, we will eventually reach a time when the billions of information exchanges shuttling through the networks at any one time would create patterns of coherence in the global brain similar to those found in the human brain. Gaia would then awaken and become her form of conscious.[74]

Optimistically, Russell concludes that "it could possibly happen within a few decades."[75] In Russell's decidedly Utopian vision of the emergence of a global brain, the creation of the Gaiafield would see the birth of a fully formed consciousness that would operate not unlike our own. (We can infer this from his use of the human brain as a referent in the selection above.)

H. G. Wells is widely credited with introducing the idea of a global brain in his 1937 essay, "World Brain: The Idea of a Permanent World Encyclopaedia." In it, Wells writes that the storage of humanity's literary legacy on microfilm

> is a fact of tremendous significance. It foreshadows a real intellectual unification of our race. The whole human memory can be, and probably in a short time will be, made accessible to every individual. And what is also of very great importance in this uncertain world where destruction becomes continually more frequent and unpredictable, is this, that photography affords now every facility for multiplying duplicates of this—which we may call?—this new all-human cerebrum. It need not be concentrated in any one single place. It need not be vulnerable as a human head or a human heart is vulnerable. It can be reproduced exactly and fully, in Peru, China, Iceland, Central Africa, or wherever else seems to afford an insurance against danger and interruption. It can have at once, the concentration of a craniate animal and the diffused vitality of an amoeba.[76]

In Wells's article, the world brain is, in fact, a large information repository, reflecting a belief in the mind/body split. When one reads Wells's original proposal of the concept of a world brain, it does not seem particularly revelatory (especially in light of the Internet). This is because Wells is interested, as we can see above, in a world brain that is all memory. In later iterations of the concept, though, the idea of a world brain begins to reflect a more nuanced understanding of brain: one that contains the ability to think as well as remember.

In Arthur C. Clarke's 1961 short story "Dial F for Frankenstein," a story that Tim Berners-Lee credits as inspiration for his work creating the Internet, the world brain gets an upgrade from mere cerebrum to a fully functional mind. The story opens:

> At 0150 GMT on December 1, 1975, every telephone in the world started to ring. A quarter of a billion people picked up their receivers, to listen for a few seconds with annoyance or perplexity. Those who had been awakened in the middle of the night assumed that some far-off friend was calling, over the satellite telephone network that had gone into service, with such a blaze of publicity, the day before. But there was no voice on the line; only a sound, which to many seemed like the roaring of the sea; to others, like the vibrations of harp strings in the wind. And there were many more, in that moment, who recalled a secret sound of childhood—the noise of blood pulsing through the veins, heard when a shell is cupped over the ear.[77]

A group of scientists analyzing the incident eventually realize that the event described above was the result of a global intelligence having been manufactured in the telephone switching system, after it was connected to a new satellite network. This global network attains sentience because it has a number of nodes equal to neurons in a human brain. While this image of spontaneous intelligence has since been refuted (the Internet has many more nodes than the self-aware telecommunications network in Clarke's story), the idea of a spontaneously emergent, global intelligence remains potent in science fiction and popular science discourse.

Clarke's story is interesting not only because it marks the emergence of a new kind of global brain (one that thinks as well as remembers) but also in that it inaugurates a new era of concern about malevolent global brains in science fiction. Clarke's story concludes with the observation that "it was far, far too late. For Homo sapiens, the telephone bell had tolled."[78] The new global brain has figured out that the scientists are plotting to disconnect the satellite network, thereby performing a "pre-frontal lobotomy" on the global brain, and moves to stop them. The story concludes by suggesting that this global intelligence is a malevolent threat to the human race. We can see in Clarke's allegory a mirrored pessimism in many articles suggesting the Internet and the future of humanity is grim, such as Nicholas Carr's famous "Is Google Making Us Stupid?" However, what Russell misses (but what Arthur C. Clarke sees in "Dial F for Frankenstein") is that any kind of global consciousness would first go through an infancy; after all, the "villain" in Clarke's story is called "Baby."

These specific images and the Internet's general feline culture are an important totem for an emerging cosmic perspective. Numerous popular cat memes, beyond just LOLcats, document cats' often amusing responses to modern technologies. A famous early cat meme was a video of a cat stalking (and catching) a trinket hanging from a ceiling fan—along with its inevitable result. Similarly, many LOLcat images show cats appearing to snowboard or being trapped in refrigerators. There's a vast collection of cats getting stuck in boxes on YouTube. In all these examples, we find cats—barely domesticated creatures who still hunt and kill large quantities of songbirds when we are not watching them—having amusing interactions with our everyday technologies at the same time that we ourselves comically struggle to understand social media and global networks of capitalism. In other words, in very real ways, we are mirroring our own anxiety about an emerging transhumanism when we trade cat memes online.

Moreover, we could argue that the language of LOLcats is not, in fact, entirely stupid (unless one wants to label any art made by children "stupid"). Rather, the amusing spellings and odd sentence structure often deployed in the captions of these images fracture and rework the English language in the process of imagining a vocabulary for talking cats. As in Clarke's story, where the malevolent actions of the global brain are actually revealed to be growing pains ("it would start looking around, and stretching its limbs. In fact, it would start to play, like any growing baby"), LOLcats creates an image-driven language in which the "normal" conventions of English are abandoned and a kind of freewheeling language play is created.

Thinking beyond the linguistic elements of the LOLcat image macros, the images themselves represent a playful experimentation with the possibilities of online ecosystems. Due to the ubiquity of image editing software on modern computers (even to the point where one can construct a LOLcat image on a browser, without other tools or software), the LOLcat phenomenon points toward a new, global conversation occurring in a language of memes. Before one dismisses the English-centric nature of meme culture, the "Tenso" meme,[79] for instance, emerged in Brazil and spread across the globe without losing the Portuguese words used in the original form. That is to say, even if a user does know exactly what "tenso" means, tenso memes are still funny. In fact, the linguistic elements of memes are not as important as the deployment of written elements as *visual* adornment within the image. This suggests, then, that the captions in LOLcat images are as important visually as they are at conveying a linguistic meaning (suggesting that using "teh" for "the" is actually directly important to the meaning of the image).[80] In fact, we could continue to suggest that the misspellings actually re-present English, defamiliarizing English-speaking Internet users from their own language. LOLcats then are sympathetic, nonthreatening vectors of a kind of postnational identity: we produce them as a means to think beyond our understanding of ourselves as national citizens and linguistic subjects. From this perspective, we can see how the system of memes, particularly as embodied by LOLcat images, represents a kind of unprecedented, global play with the ubiquitous elements of contemporary life—namely, manipulable yet alien images and words on screens.

As a result, while Nicholas Carr might suggest that making LOLCat images instead of reading Shakespeare is a sign that we are, in fact, being made stupid by the Internet, the conclusion to draw here, from an evolutionary context, is that the Internet is channeling human evolution toward different ends. As evolutionary futurists like Peter Russell and Pierre Teilhard de Chardin suggest, the increasing ubiquity of sign-circulating "new media" technologies demands the evolution of new human patterns of being. While these kinds of emerging, global consciousnesses may be obsessed at the moment with the cute and the fuzzy, this is only to be expected as humanity becomes more and more familiar with the new toys in its unfamiliar global crib. We as a noösphere are learning to deploy innate pattern recognition and combinatoric skills within the strange, contemporary contexts of a global information network.

While Web 2.0 is primarily debated as a commercial phenomenon,[81] the "crowd" that is sourced for value does not have to be commercial. All Web 2.0 is trying to do is profit from phenomena native to the Internet, as Tim O'Reilly has argued.[82] In other words, crowds were sourcing LOLcats before they were making money for companies like Facebook. Web 2.0 can thus be seen as an epiphenomenon, something that feeds off an already naturally occurring event. Fredric Jameson, glossing Hegel's concept of "Absolute Spirit" in *The Hegel Variations*, offers the following account of technology that may illuminate this phenomenon:

> What may well prove more congenial to a contemporary or a postmodern public is the invocation of Marx's notion of "General Intellect" . . . [which] evokes an historically new kind of general literacy in the mass public, most strikingly evinced in the trickling down of scientific knowledge (and technological know-how) in the population at large, a transformation that might also be described in terms of the displacement of a peasant (or feudal) mentality by a more general urban one (and in hindsight also comprehensible as a fundamental consequence of literacy and mass culture). At any rate, the hypothesis of such a social transformation in consciousness and mentality . . . strengthens the renewed appeal of Hegel's work and the revival of interest in it, in a postmodernity characterized by cynical reason and by what I will later on term plebeianization.[83]

For Marx, General Intellect was a way of concretizing Hegel's quasi-religious "Zeitgeist," the spirit of the age, in order to suggest the way that advances in technology and the mode of production manufacture new modes of being. Jameson's specific connection of General Intellect to "technological know-how" and to literacy in Marx's discussion of the birth of the urban mindset is most suggestive for thinking about Human 2.0 in the age of LOLcats. As we have seen, these memes represent a transformation in both language and subjectivity online. This transformation is both mandated by the interruptive nature of online textuality and vehiculated through the increasing ubiquity of highly advanced computing technology.

Far from the cute distraction they often appear to offer, I suggest that LOLcats are in fact the chief symptom of an emerging global intelligence. The global participatory hypertext of Web 2.0 has revealed that, as proto–Human 2.0 beings, we are increasingly enmeshed in a web of textuality that defies our ability to understand it. LOLcats are an attempt to understand this web and these new texts, tracers mapping the synapses of a new global brain. In imagining a new language for cats trapped in a modernity they do not understand, we mirror our own feelings of confusion in a world that is becoming increasingly intertwined. As our understanding of ourselves as autonomous subjects dissolves into an understanding of Earth as a single, global brain, we produce images of cats tortured by the everyday objects of modern life—fridges, computers, mirrors, furniture, snowboards—in order to capture and displace this feeling onto our pets. Exchanging images of talking cats, we are exchanging images of ourselves. Like a cat trapped in a human's world, we must come to terms with Human 2.0.

In their differing ways, the New Aesthetic and cat memes reveal the extent to which Natasha Vita-More's plea for a version of transhuman aesthetic practice, similar to the procedural architecture of Arakawa and Gins, has increasingly infiltrated mainstream modes of human expression. Both processes remind us that the future is as much an aesthetic product as it is a technological one, and both of these phenomena situate consciousness toward this extant future. On the one hand, the New Aesthetic charts an increasingly close relationship between the everyday, quotidian life and the kinds of technologies that, even twenty years ago, would have been hailed as science fictional. Meme images, meanwhile, highlight a new kind of textual future and shape the literacy practices necessary for navigating such a future. Importantly, in these two models, we find the work of evolutionary futurism being shaped and navigated through processes that are clearly aesthetic in form and in purpose. From this insight, we can conclude that, as this book has been documenting in a variety of ways, there exists a major role for textual and aesthetic humanities processes in the creation of an evolutionary future.

CONCLUSION

Acceleration and Evolutionary Futurist Utopian Practice

During the years I spent assembling this manuscript, I was continually struck by social thought's apparent allergy to evolutionary futurist *topoi.* Where, as we have seen throughout this book, evolutionary futurism is a central rhetorical mode of twentieth-century artistic and philosophical production, the rhetoric of an evolutionary overcoming has not caught on in social thought or critical theory. Partly, as I addressed in the introduction, this is due to the seductiveness of the figure of the posthuman. For all of the rhetoric of overcoming the human—of becoming the lightning bolt that issues from the cloud of man, as Nietzsche put it—evolutionary futurism, to those trained in a certain tradition of theoretical posthumanism, will always appear to fetishize the human. Within the Foucauldian, Haylesian, Wolfean tradition of posthuman thought, the human is a philosophical construct born of the excesses of Enlightenment (misogynistic, racist, rationalist) violence and one that, as a result, must be done away with.

Yet abandoning the human, which is after all the foundation of many ethical traditions, is a dubious operation in the face of a rising tide of political, economic, and ecological dehumanization. As I have been trying to argue in this volume, evolutionary futurism, understood as entangled with but ultimately distinct from contemporary transhumanism, offers a way of doing Utopia in the present that works through the humanist tradition to evolve it rather than against this tradition to displace it. At the end of the day, I find this approach to Utopia more practical than any program posthumanism might produce if it were ever to organize a Utopian strategy beyond mere negation. Partly, as I have documented in this book, while evolutionary futurism's contemporary instantiation in transhumanism suffers from a variety of problems (mostly stemming from positions that have not been sufficiently thought through), transhumanism offers a Utopian goal for humanity: the precise goal that posthuman critical theory, focused solely on the dispersion of agency and the dissolution of all human institutions, so sorely lacks. At some point in the work of negation, we must find a theory of the positive, something to sift out of the ashes of the Enlightenment.

This brings me back to a curious pattern that has struck me since the beginning of this project: the seeming lack of engagement in critical theory with the *topoi* of evolutionary futurism. As this book was coming together, however, a new thread in theory called Accelerationism emerged. Although this concept has a fairly lengthy history, the version suggested in Alex Williams and Nick Srnicek's "#Accelerate: A Manifesto for Accelerationist Politics" (2013) outlines the first attempt at an evolutionary futurist politics. In this conclusion, I briefly outline the longer history of the term "accelerationism," suggest how contemporary transhumanism already imagines what I call an "uncritical accelerationism," before turning to the 2013 manifesto and another, Laboria Cuboniks's "Xenofeminist Manifesto," as emergent attempts to imagine a contemporary evolutionary futurist political practice and critical theory. In doing so, I conclude this book about the Utopian potentials of evolutionary futurism by tracing a path from the historical archive I have been building to contemporary Utopian programs as a way of imagining an acceleration beyond the merely human.

A Brief History of a Future: The Origins of Accelerationism

To define it simply, accelerationism is the belief or tendency in Marxist thought to argue that the only way out of capitalism is *through* it, by accelerating capitalism's tendencies to break its current structures in its quest for new modes of production and new markets. Coined by Benjamin Noys in *The Persistence of the Negative,* the term emerges from his reading of post-1968 French philosophy. As Noys explains, "While many on the left responded to the rapid ebbing of the events of May with calls to Maoist or Leninist discipline, others argued the need to pursue the quasi-anarchist path of liberation from *all* structures of discipline."[1] The core canon of this other path of liberation derives from Gilles Deleuze and Félix Guattari's *Anti-Oedipus,* Jean-François Lyotard's *Libidinal Economy,* and Jean Baudrillard's *Symbolic Exchange and Death.* For Noys,

> each [author] tries to outdo the other in terms of their radicalism. In particular they reply to Marx's contention that "[t]he *real barrier* of capitalist production is *capital itself*," by arguing that we must crash through this barrier by turning capitalism against itself . . . if capitalism generates its own forces of dissolution then the necessity is to radicalize capitalism itself: the worse the better. We can call this tendency *accelerationism*.[2]

For Noys, the story mostly ends in the failure of this tendency. Noys elaborates, "Accelerationism . . . risks restoring the most teleological forms of Second International Marxism. The slogan of Bernstein's revisionism was 'the ultimate aim of socialism is nothing, but the movement is everything'; the accelerationists put a twist on this: the movement would *achieve* the aim."[3] This naive belief in capitalist movement as inherently radical, of course, "left this orientation high-and-dry when capitalism counter-attacked in the purity of its own desire for accumulation."[4] In other words, as Noys makes clear, this form of accelerationism "could offer a critique of the codified normative orderings of welfare or Fordist capitalism," but the rise of a "purer" form of capitalism (i.e., neoliberal finance capital) showed that acceleration in accumulation would not yield to liberation. In this way, Noys argues that later works of this group of philosophers (such as Deleuze and Guattari's *A Thousand Plateaus* [1980] and Baudrillard's *The Transparency of Evil* [1990]) are explicitly less radical works in order to correct or overcome or route around the early 1970s accelerationist moment. (Lyotard would take to referring to *Libidinal Economy* as his "evil book."[5]) In the face of finance capital, Noys concludes, this generation of accelerationist thinkers abandoned their original insight.

While Noys ends his account of accelerationism in *The Persistence of the Negative* with this moment of collapse in the face of neoliberalism, as he documents in *Malign Velocities* (2014), the accelerationist position does not simply die out. In the 1990s at Warwick University, the Cybercultures Research Unit (CCRU) organized by Nick Land and Sadie Plant developed a form of accelerationist practice. The CCRU adapted the Fordist libidinal liberation of Lyotard, Baudrillard, and Deleuze and Guattari to the cyberpunk moment of early Internet culture, cybernetic economics, and globalized mobile capital; as Noys explains, "Land and his colleagues at the University of Warwick strove for a new posthuman state beyond any form of the subject, excepting the delirious process of capital itself. They claimed that the replication and reinforcement of capital's processes of deterritorialization—of flux and flow—would lead to a cybernetic offensive capital could no longer control."[6] In Noys's reading, the CCRU, especially Land's version of its program, sought to accelerate the miseries of finance capital and then somehow live among the ruins. For Noys, this desire comes to suggest a form of accelerationism he labels "terminal acceleration," which is obsessed with death and destruction without offering an alternative. In many ways, the negative critical work of posthumanism is a watered-down version of the terrain CCRU worked on in the 1990s.

Terminal acceleration is important for understanding Noys's position on accelerationism. Nascent in *Persistence of the Negative* and fully developed in *Malign Velocities,* Noys's perspective regards the idea of working through capitalism as a way beyond it to be at best naive and at worst dangerous, and the position he takes results from the lingering creepiness of Nick Land's oeuvre. Despite this aura of danger (or perhaps because of it), Land's work on accelerationism was revived, according to Noys's history of the concept, following the 2008 global economic collapse. This new interest in accelerationism emerges in a number of thinkers associated with the CCRU (most notably the experimental novelist Reza Negastrani) and fellow thinkers in the British academy. This new generation of accelerationists once again (as Noys documented above) responds to Marx's idea that the limit to capitalism is capital itself, but do so, unlike in the post-1968 and cyberpunk moments, from a position in which capitalism appears to be on the brink of breaking down due to a variety of limits: spatial, temporal, computational, and ecological. This revision of accelerationism recognizes that capitalist-inflected technoscience may be the only way out of the various coming crises created by human labor for the past two to three hundred years. As McKenzie Wark writes in *Molecular Red,* in this era of constant crisis, the fact that carbon-burning industry

> is changing the climate is a knowledge that can only be created via a techno-scientific apparatus so extensive that it is now an entire planetary infrastructure. To reject techno-science altogether is to reject the means of knowing about metabolic rift. We are cyborgs, making a cyborg planet with cyborg weather, a crazed, unstable disingression, whose information and energy systems are out of joint. It's a de-natured nature without ecology.[7]

Wark follows Donna Haraway in arguing that the only possible pathway to Utopia in the present is through an engagement with the technoscientific world birthed by capitalist endeavors. There is no retreat into an imagined, preindustrial nature. That world is gone.

In this cyborg world, we need a theory that speaks to this moment—not necessarily a holism that disintegrates the human as a *mea culpa* for our past violence against the world, but a theory and a politics that meets the world we made and values the human as a key to surviving the coming chaos. As I document below, this notion of survival provides an exigence for evolutionary futurism that it has previously lacked. After all, one of the issues with many of the speculative theories of radical life extension or ecstatic becoming beyond the human is the question of why: why do we need more life? Transhumanists in the contemporary moment often answer this in terms that define death as the ultimate injustice. That answer may seem fine, but it smacks of a first-world entitlement that rightfully makes many on the left nervous. However, situated against a coming cyborg, alien Earth that no longer nurtures human life, an evolutionary futurism that agitates for a radically new humanity can meet these challenges head-on. Before elaborating on this point, though, I want to detour briefly through transhumanist thinking that touches on, though ultimately misses, accelerationist themes.

Transhumanism in the Technopresent

The futurist FM-2030, as mentioned in the introduction, inaugurated contemporary transhumanism through a series of works in the 1970s that drew on evolutionary futurist rhetoric to create a novel way of imagining that (and *acting* as though) technological change is causing humans to evolve. His work inspired more philosophically nuanced reflections on the future in Max More and other contemporary transhumanists. To read him today is to appreciate the difficult task these transhuman philosophers had in extrapolating their serious philosophy of human evolutionary *telos* from the "gee whiz" futurist ethos (similar to Hugo Gernsback) found in FM-2030's works of early transhuman thought. This disconnect between early and later transhuman philosophy results from FM-2030s stated goal: he claims human beings must adopt a philosophy of extreme positivity in order to better work within what he calls the "first age of optimism."[8] FM-2030 articulates a philosophy in which "we say Bravo to the human spirit."[9] Writing in the early 1970s, moreover, he articulates transhumanism as a celebration of the just-emerging neoliberal, globalized political and economic agenda, in addition to the more standard evolutionary futurist argument that "this very day we are on our way to a post-organic post-human stage in evolution."[10] Looking at the dawn of a neoliberal economy, FM-2030 optimistically greeted it; reading him today is hard because, with hindsight, we can see the horrific consequences he is not positioned to imagine.

FM-2030 asserts that humanity should confront the future with optimism because of the rapid rate at which our technology in the early 1970s appeared to be changing for the better:

> We are daily surging ahead in many areas: biology— genetics— physics— biochemistry— astronomy— medicine— surgery— fetology— communication— transportation— food production— computation— weather-forecasting— environmental monitoring— international relations— interpersonal relations— self-image.
>
> Advances in these and many other areas since 1955 have been more monumental than all the progress in the previous two thousand years. Even fifteen years ago, many of today's breakthroughs would have been dismissed as fantasies—too Utopian and optimistic. To us they are already routine.
>
> This rate of advance is now accelerating. Progress is faster and more global than ever.[11]

Reacting to the postwar technological and economic boom, FM-2030 extrapolated from this seeming miracle to conclude that our future was nothing but onward and upward, a trajectory Raymond Kurzweil would recycle in his works during another economic boom in the early 2000s. Although we can read FM-2030's philosophy of radical optimism as a bulwark against threats of nuclear annihilation and fears of the economic stagnation that came to pass in the late 1970s, it is also important to consider that the very technological factors he praises as signs of our optimistic future are the precursors of the mess we are now living in. Advances in food production have polluted our ground water and eroded our topsoil, environmental monitoring and weather forecasting only return increasingly dire projections, and international relations have veered increasingly toward the fundamentalism that reshaped FM-2030's native Iran.

FM-2030 is shockingly good at predicting coming technologies but often very wrong in anticipating their outcomes. In *Up-Wingers* (1973), for instance, FM-2030 celebrates the imminent emergence of a society of leisure and abundance through the creation of "cybernated" economies:

—Cybernated economies also lead to the steady obsolescence of cash-money and the rapid emergence of credit systems. Automated-magnetized-global-credit-systems enable the individual or the corporation to make small or extensive transactions anywhere on the planet without transferring or even carrying any cash at all.

—Cybernation also accelerates the rise of multinational corporations and multinational staffs.

—The waning of national economies and the continued development of regional continental and universal economies.[12]

After the 2008 housing collapse in the United States, one wonders about the optimism behind a credit-based economy and, especially, the rise of multinational corporations, whose extrajuridical, nonlocal authority have reshaped notions of justice and peace for much of the world. Continuing his surprising accuracy regarding the superficial appearance of technological and social changes, along with his characteristic inaccuracy as to their meaning, FM-2030 suggests that,

> Cybernated economies also spread abundance based not on exclusive possession but on temporary usage. People need not own but briefly rent houses— gardens— villas— yachts— helicopters— hovercrafts— computers . . . When they stay at a mobilia they enjoy all the existing commodities (as in resort hotels today) then leave them for others to enjoy.[13]

He imagines a particular vision of the future in which property will be held in common and that people will drift around the world, working when they need money and otherwise enjoying the fruits of the ever-increasing progress of technological change. However, as we have no doubt seen throughout the post-2008 world, these processes of "cybernated economies" have merely served to disenfranchise the vast majority of humanity, dooming them to the endless roaming FM-2030 is so optimistic about, only without quite as many rented hovercrafts.

This idea of a propertyless existence extends even to the body itself. In an observation that is retroactively ominous, FM-2030 writes that "in the year 2010, privacy will have far less meaning because shame and guilt and pathological fears will have greatly diminished. It will not matter to people if their conversation is overheard, their finances publicly disclosed, or their lovemaking watched."[14] FM-2030 imagines a world without shame, in which there is no privacy and therefore nothing that is not out in the open for all to see. In this new "Age of Breakthrough . . . we expect miracles because we now know that we can accomplish miracles."[15] For FM-2030, the "post-organic post-human stage in evolution" will be one of absolute freedom from social machines that force us to act in certain ways for the greater good.[16] Addressing questions of individuality in the society he envisions, he asks

> How could people lose individuality when they never enjoyed individuality in the first place?
>
> Psychologically people were owned by their authoritarian parents, clan, and church. Economically by feudal landowners and employers. Politically by the tribe, the state, the absolute ruler. In the pre-industrial world, it was not just the slave or serf who was owned but every member of society.[17]

We find in this vision a society of absolutely autonomous individuals who are, simultaneously, completely beholden to a dubiously managed commons for their basic necessities. In many ways, FM-2030's vision sounds like the contemporary West: a society of independent contractors promised and owed nothing by a crumbling welfare state but propped up by food production and resource extraction that is shifted to an unseen and increasingly brutalized neocolonial Other.

Just as his vision of the world is propped up by the unseen exploitation of others, FM-2030's entire philosophical enterprise is sustained by certain beliefs that, more than his own word "optimism," start to feel like faith—a word that transhumanism seeks to avoid. Of course, FM-2030 is probably unaware of his own faith in concepts like the human and the present, writing as he is on the other side of us from cyberpunk. As Ray Brassier and Robin Mackay explain in their introduction to Nick Land's collected writings, *Fanged Noumena,* Land finds in William Gibson's *Neuromancer* "an astonishingly complete analog for the theoretical machinery he has developed."[18] They argue that, following the arrival of a cyberpunk "textual machine for affecting reality by intensifying the anticipation of its future," Land's work moves from an intense engagement with the raw reality of *A Thousand Plateaus* into an increasingly focused exploration of cyberculture and its transformative processes.[19] From this encounter, the terminal acceleration Noys discovers through Land becomes most fully realized. Land's work following this exploration with cyberpunk increasingly takes on the character of a radical becoming-capital-instrument, in which humanity's only hope of transcendence is to merge with the global flows of capital itself. In a 2007 blog post reprinted in *Fanged Noumena,*

Land articulates this vision in a rebuttal to the tendency on the left to think of capitalism in terms of misery:

> Capitalism is still accelerating, even though it has already realized novelties beyond any previous human imagining. After all, what is human imagination? It is a relatively paltry thing, merely a sub-product of the neural activity of a species of terrestrial primate. Capitalism, in contrast, has no external limit, it has consumed life and biological intelligence to create a new life and a new plane of intelligence, vast beyond human anticipation.[20]

For Land, the human is a drag on the process of capitalism, a creation that now exceeds its creators' abilities of comprehension. In a dark revision of Teilhard, it is not the noösphere but an oeconomosphere that seems to exceed us. I draw this distinction to point out how similar FM-2030's vision is to Land's terminal acceleration, despite one being a philosophy for saying goodbye to lives spent "not playing enough not living enough not growing"[21] and the other a vision in which the sky is always "the color of television, tuned to a dead channel," as William Gibson figures it in *Neuromancer*.[22] FM-2030 correctly envisions a world of capitalism transcending Fordist limits but does not grasp that the resulting world is not made *for* humans.

Beyond placing faith in the human, FM-2030 places faith in the present, despite his assertions about optimism toward the future. In *Up-Wingers,* articulating his vision of a future so accelerated that technology has delivered miracles for us and will continue to do so, FM-2030 admonishes that "you must make a break from the traditional concept of linear historical progress."[23] In *Optimism One,* he similarly declares that "the greatest tribute to the past is to outgrow it."[24] Both cases offer a vision of the future as distinct from the past; however, we can also suggest that such a vision of time also requires the present to be a moment of revolution, thereby decoupling the past from the future. This decoupling of time is what inspires the language of miracles that peppers his writing. Technology, for FM-2030, accelerates and delivers more miracles, thereby blessing the present as the most important moment in history, this, the "first age of optimism."[25]

This conception of time, in which everything is happening now and everything that has already happened is leading up to now, is a concept Donna Haraway calls "the technopresent." In her discussion of companion species and the deep enmeshing between human and animal along the course of biological evolution, she suggests that the contemporary world, in which nothing can now be called "natural," is "technonatural biosocial modernity." In "Cyborgs to Companion Species," she writes,

> this modernity is a living fictional territory; it is always here and now, in the technopresent. With reference to anthropology's late and little-lamented "ethnographic present," the technopresent names the kind of time I experience inside the *New York Times* Science Tuesday section and on the front pages and business pages so attuned to the animation and cessation of NASDAQ. History in the technopresent . . . is reduced to the vehicle of getting to the technopresent. In the technopresent, beginnings and endings implode, such that the eternal here and now energetically emerges as a gravity well to warp all subjects and objects in its domain.[26]

Here, the technopresent speaks to the general experience of time within a scientifically structured, cybernetic society. In her earlier "A Cyborg Manifesto," as a means of resisting this conception of time, Haraway positions the cyborg as a conceptual persona for escaping not only from the trap of the technopresent but also from the older rhetorical trope that structures it: salvation history. In the 1980s moment that birthed the "Manifesto," this salvation history trope was tied specifically to nuclear war between the USSR and the United States. Haraway imagines the cyborg as a figure outside of the technopresent that is capable of countering narratives of the inevitability of nuclear war. In general, however, in becoming cyborgs, we move out of the conception of time in which the present is the only moment that matters. For the cyborg, today is *not* the perfect culmination of a past of hard work and the gateway to a future of even greater miracles.

Haraway also discusses the idea of the technopresent at length in *When Species Meet* (2007). Haraway writes that, to prophets of the technopresent such as FM-2030,

> A peculiar attitude to history characterizes those who live in the timescape of the technopresent. They (we?) tend to describe everything as new, as revolutionary, as future oriented, as a solution to problems of the past. The arrogance and ignorance of this attitude hardly need comment. So much is made to appear "new" in technoculture, linked to "revolutions" such as those in genetics and informatics. Getting through the day in technoculture is impossible without witnessing some old stability wobble and some new category make its claim on us.[27]

All these revolutions, along with the shock of the new they carry with them, which FM-2030 celebrates as grounds for a radical and new philosophy of optimism, are precisely what Haraway reminds us to be cautious about. As she writes, revolutions in the technopresent "are mostly hype."[28] In contrast to FM-2030's hyping of technological wonders, Haraway draws out the evolutionary futurism at work in our present: the increased intermingling of human and nonhuman being. She continues from this observation of technological hype and ontological reality by offering an alternative mode of thought that responds more appropriately to our moment:

> Categories abound in technocultural worlds that did not exist before; these categories are the sedimentations of processual relationships that matter. Emergents require attention to process, relationship, context, history, possibility, and conditions for flourishing. Emergents are about the apparatuses of emergence, themselves braided of heterogeneous actors and action in torqued relationship.[29]

Extending the work on the cyborg, Haraway offers this concept of the emergent—the network of actors and actions connected through mutual cohabitation on an alien Earth—as the new conceptual figure for our age.

Moreover, in decoupling the technopresent futurism of FM-2030's gadgetry from his evolutionary futurist accounts of a "post-organic posthuman stage in evolution," Haraway shows us that the way forward is not through acceleration of gadgetry but through changes in the very nature of human being, as I have been arguing throughout this book.[30] Given this relationship between technopresent, mutated ways of being, and the evolutionary futurist rhetoric that I have elucidated here, the Harawayan roots of this project are now made clear. Through the various conceptual personae Haraway gives us to resist the technopresent, she articulates a potent vocabulary for thinking through the growing disjunction between technological change and the ways in which these technologies imply but do not guarantee the creation of new forms of kinship and mutant ways of being on the Earth. In the next section, I want to turn from Haraway to two emerging political projects that take up her call to reconsider the relationship between technological change and political ways of being in the present, all through a reimagining of the tropes and methods of earlier accelerationist projects.

Evolutionary Futurist Politics against the Technopresent

In a moment dominated by the technopresent, a new generation of critical thinkers and political activists has begun to imagine an everyday Utopian practice that integrates the insights into technoscience inaugurated by Haraway's "A Cyborg Manifesto." As I mentioned previously, McKenzie Wark in *Molecular Red* suggests that we need the cyborg as a Utopian figure now more than ever, because the time for extracting ourselves from techno-modernity as a radical act has passed, if it ever existed at all. Instead, Wark suggests that Haraway is important for helping us confront the conditions of our existence *as they are* and building pathways toward other worlds from that reality. Summoning this idea of a cyborg Utopia, Wark asks:

> Can a concept of labor include scientific labor, reproductive labor, affective labor, precarious labor, even non-labor? What prevents twenty-first century labor from acting collectively by and for itself? Is it not just another kind of fetish to try to think the worker as something apart from the mesh of flesh and tech that is the composite material of the twenty-first century organization?[31]

Wark finds in Haraway's cyborg approach to science, feminism, and Utopia answers to all of these questions. Moreover, as I show in this section, two recent political manifestos—one reviving accelerationism and the other outlining a program of xenofeminism—both rely on Haraway's insights into the nature of the cyborg to build what I argue is an actual, existing evolutionary futurist Utopian program for the present. These two projects both build on the themes I have been developing in this book and point toward a future for Utopia that focuses on overcoming the human as a revolutionary act.

In both of these examples, the practice of making the future is discussed in terms of low theory, a practice that is key to moving beyond our moment. Wark explains, "Rather than imagine theory as a policing faculty flying high as a drone over all the others, lower theory is interstitial, its labor communicative rather than controlling."[32] Low theory, according to Wark, draws from forms of practice invoked by specific labor movements and, at best, "detects those emerging in key situations and alerts each field to the agendas of the other."[33] Wark's use of low theory as a description of radical thought in post-Revolution Russia and late-twentieth century California in *Molecular Red,* however, moves the insights produced by labor for us from the factory to the laboratory or the start-up coworking space. As he draws from Donna Haraway, the labor of the present is technoscientific just as it was Fordist during labor movements of the previous century.

This engagement with labor practices of technoscience is one of the most creative and important features of both evolutionary futurism as a transhistorical practice and the contemporary transhuman movement. While we can find a variety of faults with the ways these philosophies are codified, the work of transhuman thinkers such as Max More, Nick Bostrom, and David Pearce marks an important low theory engagement with technoscientific production as well as a warning about the dangers of absorbing too fully biases nascent in a field of labor. In this way, the articulation of a transhuman philosophy in the present can be seen as a means of articulating a low theory for a cybernetic future. In the two examples below, we find similar low theory engagements but ones that absorb the more maximal insights of evolutionary futurism as discussed in this book. If contemporary transhumanism merely imagines a faster version of neoliberalism, accelerationism and xenofeminism imagine ways of evolving the human into something truly alien.

Accelerationism: "The Future Must Be Cracked Open Once Again"

In 2013, the publication of Alex Williams and Nick Srnicek's "#Accelerate: Manifesto for an Accelerationist Politics" caused a minor stir online when their provocation to imagine a new collective future called for the revival of Benjamin Noys's term for the failed and dangerous politics found in the post-1968 thinkers and Nick Land, as discussed above. Rather than continue the failed libidinal revolution of the post-1968 French generation or engage in the frankly creepy cyberphilia of Land's CCRU group, this manifesto announced a new direction for the Utopian imaginary. Like the earlier accelerationists, Williams and Srnicek suggest that the way out of a stagnant capitalism in a collapsing ecosystem is through capitalism, thereby avoiding fantasies of a neoprimitive back-to-nature movement or of a "craft" ethos that will somehow slow the ever-accelerating death spiral of our climate. Unlike the post-1968 French thinkers or Land, Williams and Srnicek explicitly engage accelerationism with evolutionary futurism to solve the problems of neoliberalism and terminal acceleration.

In the opening paragraphs, Williams and Srnicek assert that, for our current model of capitalism, "the future has been cancelled": an increasingly common litany in a variety of fields,[34] due to climate change, resource depletion and, as they note, "the secular crisis of capitalism" in which "increasing automation in production processes [renders] it incapable of maintaining current standards of living for even the former middle classes of the global north."[35] In addition to these crises, there has been no corresponding emergence of a vision for the future in response to increasingly accelerating neoliberal policies of depletion and depredation. Thus, these two factors, the crisis of capitalist futurity and the inability to imagine an alternative, create the need for a new kind of Utopian imaginary. They suggest this process of imagination "entails a recovery of lost possible futures, and indeed the recovery of the future as such."[36]

To accomplish this specific vision for an accelerationist politics, Williams and Srnicek differentiate their vision from Nick Land's. In their reading, Land's vision of humanity merging with the algorithmic processes of finance capitalism merely offers "a myopic yet hypnotizing belief that capitalist speed alone could generate a global transition towards unparalleled technological singularity."[37] Demystifying Land's faith, Williams and Srnicek continue: "Landian neoliberalism confuses speed with acceleration. We may be moving fast, but only within a strictly defined set of capitalist parameters that themselves never waver."[38] Land's vision of the future is terminal speed, moving faster and faster as we rapidly fall from the sky toward the ground. Williams and Srnicek instead envision a politics that takes the "way out is through" ideas of Land and the post-1968 French accelerationist texts but couples them to "an experimental process of discovery within a universal space of possibility."[39]

This discussion of Land and the difference between speed and acceleration highlights that radical politics in the present needs to be about imagining new futures. Within neoliberalism, "we may be moving faster, but only within a strictly defined set of capitalist parameters that themselves never waver."[40] As they remind their readers, Karl Marx—unlike contemporary political movements focused on disentanglement and the creation of local pockets of anticapitalist slow culture—"was not a thinker who resisted modernity" but one who recognized that "capitalism remained the most advanced economic system to date" and remained so until neoliberalism emerged not as a "necessary historical development" but as a "contingent means to ward off the crisis of value that emerged in the 1970s."[41] For Williams and Srnicek, trying to fashion temporary pockets of resistance and valorize artisanal production ignores the fact that Marx himself was not attempting to back away from capitalist tendencies. Instead, Marx's thought was to, as Williams and Srnicek reiterate throughout their manifesto, work through capitalism's tendencies toward acceleration in order to create a future in which acceleration becomes uncoupled from capitalism.

They cite, as an example of this uncoupling, the moment in 1930 when "[John Maynard] Keynes forecast a capitalist future where individuals would have their work reduced to three hours a day."[42] Instead, they observe that we can find in the present a "progressive elimination of the work-life distinction, with work coming to permeate every aspect of the emerging social factory."[43] While we have the automation technologies Keynes imagined as precursor to less daily work, we have instead arrived at an ever-lengthening work day. Thus, they conclude that "capitalism has begun to constrain the productive forces of technology, or at least, direct them towards needlessly narrow ends."[44] At this moment, our technoscientific acceleration, the rush that makes up the core of the evolutionary futurist narrative, is beginning to decouple from the aims and goals of capitalism. It has begun to move elsewhere, but instead of progressing into those forms, we find that capitalist social organization increasingly resists the emergence of new social forms.

This resistance emerges from, in Williams and Srnicek's narrative, the left's naivety—more specifically, the complicity of certain strains of leftism with the rise of neoliberalism. Political futurists underestimated capitalism's ability to undermine the emergence of a future that no longer needs it. By imagining accelerationism as an uncanceling of the future and a quest for lost futures to reimagine, Williams and Srnicek remind us of the important lesson that underscored labor movements during the Industrial Revolution: the future must be struggled for and won. Even if the tendency to acceleration exists, we must "maximally embrace this suppressed accelerationist tendency" in order to create an evolutionary future.[45]

Williams and Srnicek's version of accelerationism is most clearly an evolutionary futurist politics when they discuss "patent wars and idea monopolization" as symptoms of "capital's need to move beyond competition." They offer the following statement—surely the most transhuman statement in their manifesto:

> The properly accelerative gains of neoliberalism have not led to less work or less stress. And rather than a world of space travel, future shock, and revolutionary technological potential, we exist in a time where the only thing which develops is marginally better consumer gadgetry. Relentless iterations of the same basic product sustain marginal consumer demand at the expense of human acceleration.[46]

From the idea of capitalist competition as a barrier to technological acceleration, they shift into arguments about human acceleration. What would it mean to accelerate the human? Williams and Srnicek offer several clues. As they suggest in this passage, the inability of our technology to advance outside a limited range of factors ("how many more megapixels are in the camera on the new iPhone?" for instance) is, in fact, a barrier to human expansion because it slows the mutational imperative of new technological forms. To this end, they argue for the need "to accelerate the process of technological evolution," an understanding of technological change that is deeply embedded in the tradition of evolutionary futurism discussed in this book. Additionally, they admonish potential accelerationists that we "never believe that technology will be sufficient to save us."[47] Instead, they argue for an understanding of technological change as a means to an end: "technology should be accelerated precisely because it is needed in order to win social conflicts."[48] Thus, we can see that human acceleration is part of technological acceleration, for true technological change—beyond iterative development of certain technologies within a sphere of sanctioned neoliberal activity—*must* imply a new articulation of the human.

Williams and Srnicek suggest that accelerationism "must also include recovering the dreams . . . of the quest of *homo sapiens* towards expansion beyond the limitations of the earth and our immediate bodily forms."[49] This striking assertion is Williams and Srnicek's most open engagement with contemporary transhumanism. However, their discussion of history in relation to this notion of escaping the bodily form, as well as the confines of Earth through space travel, diverge dramatically from the views of history espoused by many transhumanists. To take a popular example, Raymond Kurzweil's concept of "The Law of Accelerating Returns" suggests a future of increasingly plentiful returns from an uninterrupted flow of technological innovation. However, Kurzweil's model of this progress moves from Stone Age tools, through print, and arrives at computation, at which point the returns that accelerate are faster and more ubiquitous computing. Faster computation is not innovative, it is not a paradigm shift in the same way that the move from print to digital has been. This stagnation is also Williams and Srnicek's critique of Land. Both Kurzweil and Land mistake merely going faster for an acceleration that moves us into, as the manifesto concludes, a phase of "unfastening our horizons toward the universal possibilities of the Outside."[50] Acceleration is not a better phone camera but something as mutational as the introduction of print was to the human sensorium.

A break from the eternal return of neoliberalism is, according to Williams and Srnicek, the necessary precondition for realizing an evolutionary future. Thus, it seems that what contemporary transhumanism can best learn from accelerationism, with its agitation for a "properly alien future," is the failure of capitalist technoscientific evolution to naturally or inevitably deliver the kinds of radical changes proposed by many within transhumanism. Accelerationism suggests that this future has to be imagined as the product of processes other than the ones that got us where we are and then this future must be struggled for. The ideas put forth in this manifesto hold that capitalism has been an engine of innovation for generations, but its usefulness for delivering future innovations is limited. We have gone as far as we can with the tools we have had to date. As Williams and Srnicek remind us, in an age of "slow fragmentations towards primitivism, perpetual crisis, and planetary ecological collapse . . . the future needs to be constructed."[51]

Xenofeminism: "If Nature Is Unjust, Change Nature!"

Responding to similar issues as Williams and Srnicek, the feminist activist collective Laboria Cuboniks published their "Xenofeminist Manifesto" online in 2015. As Emma Wilson argues, Laboria Cuboniks's piece connects accelerationist impulses to a longer lineage of cyberfeminism, especially the strand of world-making ontology inaugurated by Donna Haraway in "A Cyborg Manifesto."[52] As such, in this section, I discuss xenofeminism in the context of the strong connection between evolutionary futurism and accelerationist political action. Reading xenofeminism in this way, I emphasize that this project of feminist ontology offers a strong model for enacting the evolutionary futures held out by accelerationism and documented throughout the history of evolutionary futurism explored in this book. In xenofeminism, an inclusive politics for constructing an accelerationist future is offered as a blueprint for effective action. Borrowing from the long tradition of feminist organizing and action, I want to argue that xenofeminism, even more than the fairly theoretical "Accelerationist Manifesto," offers a model for imagining a transhuman future that breaks from, as Laboria Cuboniks calls it, "futureless repetition on the treadmill of capital" (0x00).[53]

Rather than the image of the future being canceled, as deployed by Williams and Srnicek, Laboria Cuboniks takes a more immediately praxis-oriented approach to the same topic: "Our future requires depetrification," they argue (0x00). The rhetorical shift from "the future has been canceled" to "our future requires depetrification" moves accelerationist themes from theory into practice. One of Laboria Cuboniks's central demands is "no more reification of the given masked as critique" (0x00). Peter Berger and Thomas Luckmann have defined reification as "the apprehension of human phenomena as if they were things, that is, in non-human or possibly supra-human terms."[54] They continue that "reification implies that man is capable of forgetting his own authorship of the human world, and further."[55] In other words, Laboria Cuboniks implies that our notion of our own future (one of more and more intense forms of neoliberal control and increasingly dispiriting work on a dying planet) has become subject to this process: it is the result of a series of contingent human decisions that have now been made to appear inevitable. In addition to the usual subjects of reification (class, government, economy, power) our destiny now seems sadly unavoidable. Thus, Laboria Cuboniks makes a key move here to shift accelerationism toward praxis. To say that the future has been canceled and must be reinvented is to imply that there is nothing to work with and that imagination is the primary mode of political engagement. To say, instead, that the future requires depetrification, an act of taking something seemingly set in stone and showing that it is written on clay, is to immediately hail action—to focus on the work that goes into creating a more just future.

Laboria Cuboniks continues on this topic of the work of the future: "XF is not a bid for revolution, but a wager on the long game of history, demanding imagination, dexterity, and persistence" (0x00). Working for the future then becomes a process whose results we may not see. Rather, xenofeminism outlines a version of history in which short-term major goals are sacrificed in favor of long-term accomplishments. This model of history is in line with a lot of the evolutionary futurist rhetoric we have seen in this book, but it is intriguingly also incompatible with certain versions of contemporary transhumanism. Contemporary transhumanism promises a future of immortality and extropian delights for its participants, available within their lifetime. Instead, xenofeminism offers a political program for working for a future that may never arrive but will, instead, pay increasing dividends to all, not just those involved in the work.

Given that this work involves reversing the reification of the future, xenofeminism might seem coincident with the long history of consciousness raising in both feminist and Marxist praxis. In "forgetting his own authorship of the human world," as Berger and Luckmann frame reification, humanity becomes alienated from its ability to shape and reshape the social conditions of its existence.[56] In such a model of politics, the task of the activist is to make subjects enmeshed in the alienating trap of reification aware of this entanglement, which serves as a precursor to the emergence of a new political consciousness. However, Laboria Cuboniks articulates xenofeminism as the opposite: rather than freeing people from their own alienation, xenofeminism's "construction of freedom involves not less but more alienation; alienation is the labour of freedom's construction" (0x01). As they articulate, we live in an age in which "nothing should be accepted as fixed, permanent, or 'given'—neither material conditions nor social forms" (0x01). Through this, they are able to hail as allies

> anyone who's been deemed "unnatural" in the face of reigning biological norms, anyone who's experienced injustices wrought in the name of natural order, will realize that the glorification of "nature" has nothing to offer us—the queer and trans among us, the differently-abled, as well as those who have suffered discrimination due to pregnancy or duties connected to child-rearing. (0x01)

By summoning a range of bodies and experiences as within the purview of this movement, Laboria Cuboniks positions xenofeminism as being against nature and against any notion of naturalism.

Like many in contemporary transhumanism, Laboria Cuboniks connects this need to abandon naturalist ideas to the fettering of technology in the present. A common argument for transhumanism, which we most notably encountered in David Pearce (although it also runs through much transhuman thinking), is that the belief in the sanctity of the human form holds back technological experiments with our own forms. However, this argument often emerges from cis-gendered, white males who want to move beyond the limits of their privilege. In xenofeminism, this desire to move beyond is tied to the idea that discourses of "the natural" already alienate a wide range of bodies from themselves, even without the help of technological intervention. Rather than extend notions of white male cis privilege further into a technoscientific evolution, xenofeminism stands as a challenge to become even more alienated.

With an eye toward increasing our becoming alien as an evolutionary, political project, Laboria Cuboniks echoes the call to unfetter technological advance from capitalist market imperatives (0x03). Beyond suggesting that technology's "rapid growth is offset by gloat," they move to a broader project: arguing for the need to recoup "rationality" as a feminist political strategy (0x03). Rather than continue the historical project of mapping and documenting patriarchy within the discourses of science and technology and even beyond the idea of articulating a feminine rationality, Laboria Cuboniks instead argues that

> Science is not an expression but a suspension of gender. If today it is dominated by masculine egos, then it is at odds with itself—and this contradiction can be leveraged. Reason, like information, wants to be free, and patriarchy cannot give it freedom. *Rationalism must itself be a feminism.* (0x04)

Instead of imagining reason as a tool for dominating nonwhite, female, and other othered bodies, Laboria Cuboniks suggests instead that reason can be "an engine for feminist emancipation" and a powerful tool for "everyone to speak as no one in particular" (0x04). Later in the manifesto, they connect this idea of speaking as no one in particular to the specifically cyberfeminist "potential of early, text-based internet culture for countering repressive gender regimes, generating solidarity among marginalized groups, and creating new spaces for experimentation" (0x13). In other words, when all you have is text, it is hard to police things like identity and other categories of hierarchy tied to physical embodiment. Still, "the dominance of the visual in today's online interfaces has reinstated familiar modes of identity policing, power relations and gender norms in self-representation" (0x13). As they claim throughout the manifesto, just because the text-based world that spawned the radical potentiality of cyberfeminism has passed into history does not mean that the tactics and lessons learned are therefore irrelevant.

Instead of mourning the loss of a moment for cyberfeminism—a moment Laboria Cuboniks positions as intensely alienated—the collective argues that we now require "a feminism sensitive to the insidious return of old power structures, yet savvy enough to know how to exploit the potential" (0x13). This need is connected to its argument for rationality as a feminist method, instead of "the excess of modesty in feminist agendas of recent decades[, which] is not proportionate to the monstrous complexity of our reality, a reality crosshatched with fibre-optic cables, radio and microwaves, oil and gas pipelines, aerial and shipping routes, and the unrelenting, simultaneous execution of millions of communication protocols with every passing millisecond" (0x05). Into this chaos, it positions xenofeminism as a means to a "freedom-to rather than a freedom-from," an "ideological infrastructure" that offers "more than digital self-defense and freedom from patriarchal networks" (0x07). Instead, xenofeminism is a rational alienation in which "feminists . . . equip themselves with the skills to redeploy existing technologies and invent novel cognitive and material tools in the service of common ends" (0x07). In this claim, I recognize xenofeminism as an attempt to preserve the lessons about technology and standpoint taught by figures like Donna Haraway and leverage them into a global imaginary that can counteract the sweeping scale of big data as a world-making enterprise.

Practicing world making inspires Laboria Cuboniks's claim that xenofeminism is "gender-abolitionist" (0x0E). Rather than seek for the abolition of gendered characteristics, they claim "the ambition to construct a society where traits currently assembled under the rubric of gender, no longer furnish a grid for the asymmetric operation of power" (0x0E). This interest in thinking a society outside of a grid for power focused on gender (they also mention race as another grid xenofeminism seeks to overcome) highlights their inspiration in text-based cyberfeminist agitation, as well as their commitment to rationality as a feminist practice. If reason is, as they claim, a discourse that seeks to speak the world without the markers of gender, such a position for knowledge production can be a powerful tool in disarticulating gender and race as grids in which power operates. In order to accomplish a worldview on this scale, Laboria Cuboniks intensifies its claim to rationality as a feminist tool: no less than "the viability of emancipatory abolitionist projects—the abolition of class, gender, and race—hinges on a profound reworking of the universal" (0X0F). Another term to which contemporary social thought has had an allergy, xenofeminism seeks a universal "built from the bottom up—or, better, laterally, opening new lines of transit across an uneven landscape" (0x0F).

This new "universal must be grasped as generic, which is to say, intersectional" (0x0F). It must be supple and open to adaptation. Borrowing from the software engineering practice of continuous delivery, in which teams work to deliver workable code quickly so that a product can be tested and released with greater speed, Laboria Cuboniks proposes that construction of this new universal "is therefore understood to be a negentropic, iterative, and continual refashioning" process of continuous delivery (0x10). Rather than hegemonic control over the construction and transmission of universals, xenofeminism is "like open source software . . . available for perpetual modification and enhancement following the navigational impulse of militant ethical reasoning" (0x10). This continuous development of new, emergent universals allies xenofeminism with the approach to transhumanism described as "extropia" by Max More. Xenofeminism imagines an evolutionary future for the human condition as structured by universal definitions that evolve along with our physical, libidinal, and psychological forms. Toward this end, they suggest that "xenofeminism is a platform, an incipient ambition to construct a new language for sexual politics—a language that seizes its own methods as materials to be reworked, and incrementally bootstraps itself into existence" (0x19). Continuing with their use of open source and other software engineering practices, they assert that xenofeminism is engaged in "constructing an entire universe of free and open source platforms that is the closest thing to a practicable communism many of us have ever seen" (0x16).

This idea of bootstrapping and evolving universals has practical as well as theoretical payoffs for Laboria Cuboniks. As an example of the hacker mindset they inaugurate as feminist practice, the xenofeminist manifesto singles out hormones as a key site for rational, universalist experimentation in gender abolition:

> Hormones hack into gender systems possessing political scope extending beyond the aesthetic calibration of individual bodies. Thought structurally, the distribution of hormones—who or what this distribution prioritizes or pathologizes—is of paramount import. The rise of the internet and the hydra of black market pharmacies it let loose—together with a publicly accessible archive of endocrinological knowhow—was instrumental in wresting control of the hormonal economy away from "gatekeeping" institutions seeking to mitigate threats to established distributions of the sexual. (0x16)

Access to hormones that reprogram the basic operations of gender can be considered one vector toward gender abolitionism. As Paul B. Preciado's *Testo Junkie* makes clear, hormones are not just for making men into women and vice versa; they are a technology for opening up an array of possible gender positions *and,* as Laboria Cuboniks emphasizes throughout the manifesto, signal the ultimately arbitrary nature of gender as a basis for a grid of asymmetrical power in the age of hackable wetware.[57] Where Williams and Srnicek imagine a general acceleration of the human as part of their accelerationist program, xenofeminism in its focus on gender hacking with hormone technology in particular and with the creation of "free and open source medicine" more generally, imagines specific and implementable goals for the hacking of the human form in an evolutionary manner as part of a program for liberation.

Moreover, this process of liberation is in tune with the general understanding of evolution as gradual and long term. Rethinking the glorification of speed in futurist rhetorics, Laboria Cuboniks writes:

> Ours is a transformation of seeping, directed subsumption rather than rapid overthrow; it is a transformation of deliberate construction, seeking to submerge the white-supremacist capitalist patriarchy in a sea of procedures that soften its shell and dismantle its defenses, so as to build a new world from the scraps. (0x19)

Theirs is the slow revolution of biosphere over geosphere in Vernadsky's work rather than an immediate rush toward a radical break. It understands that evolution is a process that can be molded and shaped—one that can be directed toward a goal of more radical equality and inclusiveness, not just toward the creation of better humans along an already-existing vision of white male power. As Laboria Cuboniks suggests, "xenofeminism indexes the desire to construct an alien future with a triumphant X on a mobile map. This X does not mark a destination" (0x1A). As Max More gestured toward in rejecting the notion of a fixed utopian goal, the alien future never fully arrives. All we have is more and more alienation from a collection of supposedly natural identities that, frankly, do not work very well anymore.

Creating more and better alienation while moving toward an alien future, Laboria Cuboniks channels the "the way out is through" ethos of accelerationism; however, by coupling this ethos to a desire to demolish gender as a grid for power *and* to the recognition that this project is a long way from complete, xenofeminism imagines something that starts to look like a plan for enacting a truly radical, truly alien future for human evolution. By moving outside the grids of gender and race and class, by imagining reason as a method uniquely suited for feminist action, and by highlighting software and platform development as the means to attain gender-abolitionist ends in the present *and* the future, xenofeminism not only extends accelerationism into feminist inquiry but, more important, documents the first fully realized set of political protocols for doing evolutionary futurism as a practice instead of just as a philosophy. Xenofeminism points the way toward an evolutionary futurist utopian practice for the present.

Our Alien Future

As a means of concluding this rhetorical history of evolutionary futurism, I turn finally to this notion of alien futurity that animates xenofeminism. If Williams and Srnicek imagine that the future is something we have to do political work to recover for the present, Laboria Cuboniks imagines a program of action for actualizing an alien future. Beyond reiterating these protocols for platform building, message saturation, gender abolition, and open source medicine, I conclude by considering what committing to a xenofeminist form of practice might involve, particularly as it might mutate contemporary transhumanism from within. As I have been arguing, accelerationism and xenofeminism represent the first attempts to seriously consider transhumanism as a powerful source of tropes for imagining an oppositional politics in the present. However, to consider further the relationship between transhumanism and these newly emergent political vectors, I want to make a potentially strange detour: specifically to Classical Athens and a consideration of Plato's *Gorgias.* In that dialogue—which tracks the relations between desire, politics, and rhetoric, as they are used to determine what is best for the citizens of a state—we can reconstitute a discourse that addresses the shared usage of the phrase "collective self-mastery," which both manifestos (in Section 03.14 of "#Accelerate" and in 0x00D in "Xenofeminism") use as a figure for evolutionary futurist political practice.

In Socrates's discussion with Gorgias, Polus, and, most important, Callicles, a vision forms of what collective self-mastery might mean to us in the present time of crisis addressed by accelerationism and xenofeminism. In Socrates, of all figures, we begin to see what it might look like to work for an alien future. Toward the end of the dialogue, Socrates's attempt to engage his friend, the tyrant Callicles, goes quite sour.[58] In this dialogue, which debates the nature of justice and the ability of rhetoric to deliver it to the world, Socrates declares:

> But, my blessed man, please see whether what's noble and what's good isn't something other than preserving and being preserved. Perhaps one who is truly a man should stop thinking about how long he will live. He should not be attached to life but should commit these concerns to the god and believe the women who say that not one single person can escape fate. He should thereupon give consideration to how he might live the part of his life still before him as well as possible.[59]

In this moment, Socrates frowns upon Callicles's construction of a democractic society as one in which the will of the orator *is* the will of the city and in which those who are better than most (a phrase Socrates, throughout the dialogue, repeatedly challenges Callicles to define) should use their stronger will to secure the best for themselves. Socrates challenges

Callicles's assertion that life is best when it is a never-ending series of sexual and gustatory encounters by instead offering the idea that the role of our lives is to direct those around us toward the best based on our impressions of that truth.[60] For Socrates, in a series of speeches in which Plato has him predict his own death and offer a justification for why he chooses to drink hemlock, the goal of a good life is not to attain more and more time on Earth but to spend the time we have doing the best for ourselves and passing along what we have learned to others. This is, precisely, the "well-examined life" Socrates so famously sought to live.

I turn from radically futurist political visions to ancient Greece because Socrates's admonition to his politically ambitious friend is apropos to what we have been considering in this conclusion—not to mention in this project as a whole. "Live long enough to live forever" is a slogan of Raymond Kurzweil, one that gets repeated in discussing contemporary transhumanism, especially the versions of it most like Kurzweil's Singularitarianism. While this version of transhumanism rhetorically positions itself on the side of Socrates, by claiming that it wants to abolish death for all, the monetary expense of their solutions and their unwillingness to invest in political paradigms outside of neoliberal capitalism highlight the fact that this slogan is much more rhetorically similar to Callicles's vision: what is best for the best is what we all have to accept as our future. Until fairly recently, this rhetorical position was dominant in transhumanism. However, as we have seen, accelerationism and xenofeminism take the tropes of evolutionary futurism as a starting point for moving beyond a world of futurological stagnation.

Unlike contemporary transhumanism's ethos of faith in the inevitability of a radical break, Williams and Srnicek and Laboria Cuboniks, like Socrates, position their political platforms as radical experiments. Just as it was for democracy itself in Classical Athens and Enlightenment Europe, the outcomes are not known in advance. In discussing the difference between an oratory based on flattery and one focused on the production of justice, Socrates characterizes the former: "through routine and knack it merely preserves the memory of what customarily happens, and that's how it also supplies its pleasures."[61] I fear that without a radical vision of otherness, one specifically beyond a capitalism that is clearly a barrier to radical technological and evolutionary change, contemporary transhumanism is destined to become the "routine and knack" of an already overdetermined futurism. Instead, as accelerationist and xenofeminist politics remind us, the future of the future must coemerge with the platforms on which we build it. The ways of the past and the present are insufficiently equipped to unlock the kinds of radical futures evolutionary futurism has imagined. Indeed, Plato was critiquing the recent past of Athenian democracy—only recently restored from the governance of the Thirty Tyrants and a series of disastrous military endeavors that greatly and permanently weakened the city-state—via Socrates's engagement with Polus and Callicles. For instance, when Socrates famously discusses the difference between art (*techne*) and knack in his discussion with Polus (462b–466a), Socrates declares the rhetoric practiced by the Sophists to be a practice of flattering the soul analogous to pastry baking,[62] a practice that flatters the body. Socrates instead exhorts Callicles to purify his own soul first; only then can he hope to break with tradition and lead the Athenians into a future that is not just "that fit of sickness" that comes from eating too much cake.[63]

In "#Accelerate," Williams and Srnicek define accelerationist politics as a mode of articulating a new notion of mastery (which I quote at length to reproduce the mechanics of their argument):

> We declare that only a Promethean politics of maximal mastery over society and its environment is capable of either dealing with global problems or achieving victory over capital. This mastery must be distinguished from that beloved of thinkers of the original Enlightenment. The clockwork universe of Laplace, so easily mastered given sufficient information, is long gone from the agenda of serious scientific understanding. But this is not to align ourselves with the tired residue of postmodernity, decrying mastery as protofascistic or authority as innately illegitimate. Instead we propose that the problems besetting our planet and our species oblige us to refurbish mastery in a newly complex guise; whilst we cannot predict the precise result of our actions, we can determine probabilistically likely ranges of outcomes.[64]

Rather than throw up their hands in the face of a universe defined by complex processes, declaring the world to be "undecidable" and mastery to be inherently corrupt, Williams and Srnicek call on the idea of imagining mastery in the form of a cybernetic model: one based on probabilities rather than certainties, and one capable of being supple enough to survive first contact with a messy reality. They cite the famous Chilean experiment in computerized economic management, Project Cybersyn, as a model for what such processes might look like.[65] Unlike many antitechnology stances in anticapitalist thought, Williams and Srnicek see in the figure of big data the possibility of using technologies of control as means for producing a more just, more fair, and more merciful society.

In "Xenofeminism," Laboria Cuboniks imagines this process slightly differently. The collective explains that

> collective self-mastery requires a hyperstitional manipulation of desire's puppet-strings, and deployment of semiotic operators over a terrain of highly networked cultural systems. The will will always be corrupted by the memes in which it traffics, but nothing prevents us from instrumentalizing this fact, and calibrating it in view of the ends it desires.[66]

"Hyperstition" here refers to a neologism coined by Nick Land's CCRU and glossed by cyberculture theorist Delphi Carstens as a term used to "describe the action of successful ideas in the arena of culture."[67] Carstens explains that hyperstitions function as "magical sigils or engineering diagrams" and "are ideas that, once 'downloaded' into the cultural mainframe, engender apocalyptic positive feedback cycles."[68] Laboria Cuboniks thus points to the work of collective self-mastery as manipulating the quasi-magical forces that compel us to act in certain ways, attempting, I argue, to treat ideology as a process of cybernetic feedback. Carstens points to CCRU's association between hyperstition and H. P. Lovecraft's Old Ones, those monstrous beings who lurk beyond human perception and compel us to act. In xenofeminist practice, the distortion and manipulation is already part of the global network of meaning circuitry, so learning to bend these processes to other ends becomes the chief political task of our day.

This notion of a meaning that comes from outside our individual consciousness resonates with the Socratic project of creating the examined life and, only then, attempting to master others. Speaking to no one in particular at the conclusion of *Gorgias,* after having repeatedly failed to engage Gorgias, Callicles, and Polus in dialogue, Socrates offers the following on what the good life might mean for a philosopher such as himself:

> Nothing terrible will happen to you if you really are an admirable and good man, one who practices excellence. And then, after we've practiced it then at last, if we think we should, we'll turn to politics, or then we'll deliberate about whatever subject we please, when we're better at deliberating than we are now. For it's a shameful thing for us, being in the condition we appear to be in at present—when we never think the same about the same subjects, the most important ones at that—to sound off as though we're somebodies.[69]

For Socrates, it is better first to examine oneself before moving into the realm of governance. As political philosopher Isaiah Berlin makes clear, this notion is antithetical to understandings of the self that dominate Classical Athens. Surveying contemporaneous ideas of democracy, Berlin concludes that "there is no note of individualism here, of the value of the State consisting in what it contributes to individual satisfaction."[70] Berlin suggests that any modern notion of individual rights was "realized only within and as part of the life of the Greek *polis*."[71] In this way, Berlin suggests in an example (which he uses frequently in his writings on liberty) that Athenians in this period resemble students at a boarding school: "the school may take pride in the fact that it does not need to threaten or bully, punish or intimidate, but it is the collective spirit of the school, the solidarity of its members, that is being praised."[72] In Socrates's Athens, the only kind of mastery imaginable was of this broader *polis*. As such, *Gorgias*, in both Polus and Callicles, is marked by characters who view the city as a means toward the satisfaction of their voracious appetites for food, sex, violence, and the like. In debating with Polus, Socrates instead asks after self-rule, "I mean each individual ruling himself. Or is there no need at all to rule himself, but only to rule others?"[73]

Thus, in *Gorgias*, with its emphasis on the willingness to die for one's beliefs and temper one's insatiable desire for more as a necessary precursor for the dominance of others, we find a possible definition for "collective self-mastery" in the accelerationist, xenofeminist sense: the ability to leverage our own corrupt tendencies and desires toward the processes of self-discovery that can moderate these same destructive tendencies in others in the future after we are gone. Collective self-mastery becomes the creation of processes for overcoming the limitations of all bodies and minds for universal transcendence against the ephemeral power bought by money or status. Rather than the endless repetitive ingestion of sex, food, and violence that are the fruits of power for Callicles, both manifestos call for an ongoing unfolding of the future as radical difference, an experimental extrapolation of what humans can do. Such a future would be truly alien, rather than just the same tired reiteration of the present.

So, maybe it is not the Singularitarian slogan, "Live long enough to live forever," that evolutionary futurism actually needs. Transhumanism's lack of political leverage and overall *weirdness* emerges from this fundamental issue: notions of future building are incompatible with the violent accumulation of power and life within individuals. Instead, tasked with creating the future through collective self-mastery in accelerationism and xenofeminism—and bearing in mind Socrates's understanding of liberty and the examined life—the Kurzweilean slogan might become "live so that the future might live forever." Perhaps living for the continued renewal of alien possibility is and always has been the core of the rhetorical position called evolutionary futurism.

ACKNOWLEDGMENTS

This book exists because of the insight and inspiration of two teachers: Rich Doyle and Jeff Nealon. Rich's wide-ranging wisdom helped shape my tangle of interests into a coherent project and pointed me toward the mystical content that pops up throughout this book. The day, sitting on folding chairs in a field in rural Pennsylvania drinking a thermos of herbal tea, when he casually asked, "why don't you write a history of transhumanism?" launched me on the trajectory that I've been following ever since. Jeff's careful instruction in reading has been of great importance to my method. Without his Fredric Jameson seminar at Penn State, this book would not have evolved into the project it is today.

Thanks to Doug Armato for believing in this project and his thoughtful editing and to Erin Warholm-Wohlenhaus for putting it all together. Additional thanks to Joni Adamson, Josh DiCaglio, and Jenell Johnson for their support during the writing and revision of this book.

This manuscript was written in LaTeX, managed with Git, and compiled into its final form with Pandoc. The communities who maintain these projects (as well as Biber, CSL, Citeproc, and Homebrew) greatly helped my writing. My Twitter followers constantly remind me that technology is, for better for worse, fundamentally altering the way we think about and interact with the world. I think differently about language and community because of you all: thank you. An alarming number of the primary texts from the early 2000s I work with have vanished from the Internet during the writing and revising of this book. Without the Internet Archive, I literally would not have been able to finish the Introduction, chapter 4, and the Conclusion. The Archive's hard work is invaluable for preserving our digital cultural heritage.

My parents, Tom and Judy Pilsch, told me to finish my computer science degree when I proposed switching to English. I am glad they did; without that, none of this would make sense. I have always been able to count on their love and support, even when they have no idea what I'm talking about.

Finally, the ultimate lesson of this book is that humans actually *are* more interesting than machines (which is not where I started from). Without Shawna Ross, I would not have arrived at this conclusion. Thank you for all your love and support. Shawna read the entire manuscript (more than once) and provided invaluable feedback as it emerged into its final form. I could not (and probably would not) do any of this without her.

NOTES

Introduction

1. The futurist born Fereidoun M. Esfandiary, whose adopted moniker indicated his desire to live to the year 2030, died in 2000.
2. "The Transhumanist FAQ 3.0. Humanity+," January 2013, http://humanityplus.org/ n.p. The original 1990 text from which this quote emerges is not available online but is still quoted in the "Transhumanism FAQ."
3. Max More, "Transhumanism: Towards a Future Philosophy. Max More," 1996, http://maxmore.com/transhum.htm n.p. This article, hosted on Max More's website since 1996, disappeared off his website in 2014, apparently accidentally. It is still accessible through Archive.org's Wayback Machine.
4. George Dyson, *Turing's Cathedral: The Origins of the Digital Universe* (New York: Pantheon, 2012), 300.
5. More, "Transhumanism," n.p.
6. Ibid.
7. The term "posthuman" (or "post-human" as FM-2030 renders it) is present in many of the foundational works of post-1970s transhuman thought. The term is not used in the same exact way it is often used in academic circles, except perhaps in N. Katherine Hayles's *How We Became Posthuman*. For many academics, posthuman speaks to the condition of being beyond humanist thought, first suggested by Michel Foucault at the end of *The Order of Things*. However, for transhumanists, the term speaks to the literal configuration of human existence after transhuman mutations.
8. I do not engage with Ettinger—the father of cryonics (the freezing of people at the moment of death in order to reanimate them in the future)—to any great extent in this book, primarily because the rhetorical and philosophical implications of Ettinger's work have been thoroughly and spectacularly documented in Richard Doyle's *Wetwares* (2003), a book that in many ways stands as the precursor to this one. See Richard Doyle, *Wetwares: Experiments in Postvital Living* (Minneapolis: University of Minnesota Press, 2003).
9. Alvin Toffler, *Future Shock* (New York: Bantam, 1984), 2.

10. Ibid.
11. Ibid.
12. Robert Ettinger, *Man into Superman: The Startling Potential for Human Evolution—and How to Be a Part of It* (Clinton Township, Mich.: Cryogenics Institute, 1989), 9, http://www.cryonics.org/images/uploads/misc/ManIntoSuperman.pdf.
13. Cary Wolfe, *What Is Posthumanism?* (Minneapolis: University of Minnesota Press, 2009), xiii.
14. Ibid., xv.
15. Ibid.
16. Specifically, see Hava Tirosh-Samuelson and Kenneth L. Mossman, eds., *Building Better Humans?: Refocusing the Debate on Transhumanism* (Frankfurt am Main: Peter Lang, 2012), and Calvin Mercer and Tracy J. Trothen, eds., *Religion and Transhumanism: The Unknown Future of Human Enhancement* (Santa Barbara, Calif.: Praeger, 2014).
17. More, "Transhumanism," n.p.
18. David F. Noble, *The Religion of Technology: The Divinity of Man and the Spirit of Invention* (New York: Knopf, 1997), 6.
19. Ibid., 12.
20. I return to this relationship between technology and metaphoric wish fulfillment in chapter 2, but it is important for rethinking how transhumanism differs from Noble's thesis, here.
21. Ker Than, "Hang in There: The 25-Year Wait for Immortality," *LiveScience.com*, November 4, 2005, http://www.livescience.com/.
22. Nicholas Agar, *Humanity's End: Why We Should Reject Radical Enhancement* (Cambridge, Mass.: MIT Press, 2010), 1.
23. Nicholas Agar, *Truly Human Enhancement: A Philosophical Defense of Limits* (Cambridge, Mass: MIT Press, 2013), 36.
24. Ibid., 36–39.
25. Max More, "A Letter to Mother Nature," 1999, http://www.maxmore.com/mother.htm.
26. Ibid.
27. John Poulakos, "Toward a Sophistic Definition of Rhetoric," *Philosophy & Rhetoric* 16, no. 1 (January 1, 1983): 36.
28. Ibid., 43.
29. Chaïm Perelman and Lucie Olbrechts-Tyteca, *The New Rhetoric* (South Bend, Ind.: University of Notre Dame Press, 1991), 154.
30. Ibid.
31. Timothy Morton, "Here Comes Everything: The Promise of Object-Oriented Ontology," *Qui Parle: Critical Humanities and Social Sciences* 19, no. 2 (2011): 171.
32. As I discuss in chapter 2, a recent special issue of *The Journal of Evolution and Technology* addressed the question of Nietzsche and transhumanism. Additionally, Keith Ansell-Pearson's *Germinal Life: The Difference and Repetition of Deleuze* (New York: Routledge, 1999) offers a transhuman reading of Deleuzian thought

that relies heavily on Deleuze's relationship to Bergson as a means of unlocking this perspective. While only two examples, they point to the strong affinity for transhuman arguments present in both thinkers' arguments.

33. Ray Kurzweil, *The Singularity Is Near: When Humans Transcend Biology* (New York: Penguin, 2006), 10.
34. John von Neumann, "The General and Logical Theory of Automata," in *Collected Works Volume V: Design of Computers, Theory of Automata and Numerical Analysis*, ed. A.H. Taub (New York: Pergamon, 1963), 289.
35. Mark Ward, *Virtual Organisms: The Startling World of Artificial Life* (New York: Macmillan, 2000), 65.
36. Based on biographical accounts of him, von Neumann was apparently a fairly disorganized genius in the classic model (famously getting multiple tickets in Princeton, N.J., for attempting to drive while reading and constantly bothering his down-the-hall neighbor at the Institute for Advanced Studies, Albert Einstein, with German march music, which he listened to at deafening volumes). This absentmindedness extends to his publication record. Arthur W. Burks compiled *Theory of Self-Reproducing Automata* from lecture transcriptions and notes von Neumann left at his death. *First Draft of a Report on the EDVAC* was meant as an internal memo but was edited and published by Andrew Goldstine. Much of what we know of von Neumann's automata work, in terms of the grand scope he envisioned, further emerges from encomiums by colleagues such as Stanislaw Ulam and Freeman Dyson.

    Nonetheless, the quote is everywhere attributed to von Neumann online and in the literature of artificial life, although I cannot find a source. However, as much of what was attributed to von Neumann's theories during this period emerges from reminiscences and interviews with his colleagues and friends, the possibly apocryphal nature of this quote is not entirely surprising.
37. Christopher G. Langton, "Artificial Life," in *Artificial Life: Proceedings of an Interdisciplinary Workshop on the Synthesis and Simulation of Living Systems*, ed. Christopher G. Langton (Redwood City, Calif.: Westview, 1989), 11.
38. John von Neumann, *Theory of Self-Reproducing Automata*, ed. Arthur W. Burks (Champaign: University of Illinois Press, 1966), 20.
39. Martine Rothblatt, "From Mind Loading to Mind Cloning: Gene to Meme to Beme; a Perspective on the Nature of Humanity," in *H+/-: Transhumanism and Its Critics*, ed. Gregory R. Hansell (Philadelphia: Metanexus Institute, 2011), 113.
40. Max More, "True Transhumanism: A Reply to Don Ihde," in *H+/-: Transhumanism and Its Critics*, ed. Gregory R. Hansell and William Grassie (Philadelphia: Metanexus Institute, 2011), 140.
41. This move would also stand as a rejoinder to Cary Wolfe, who dismisses transhumanism as being "[derived] directly from ideals of human perfectibility, rationality, and agency inherited from Renaissance humanism and the Enlightenment." Wolfe, *What Is Posthumanism?*, xiii.
42. For more on the utopian socialists and the difficulty in figuring out who they actually were, see Roger Paden, "Marx's Critique of the Utopian Socialists," *Utopian Studies* 13, no. 2 (January 1, 2002): 67–91.

43. Fredric Jameson, *Archaeologies of the Future: The Desire Called Science Fiction and Other Science Fictions* (London: Verso, 2005), 10.
44. Ibid., 227.
45. Fredric Jameson, *Valences of the Dialectic* (London: Verso, 2009), 410.
46. Jameson, *Archaeologies of the Future,* 227.
47. Ibid., 193.
48. For an extensive discussion of *Jaws* and political resistance, see Fredric Jameson, "Reification and Utopia in Mass Culture," *Social Text,* no. 1 (January 1, 1979): 130–48. For a discussion of *Body Heat,* see Fredric Jameson, *Postmodernism, or, the Cultural Logic of Late Capitalism* (Durham, N.C.: Duke University Press, 1992).
49. Fredric Jameson, *The Political Unconscious* (Ithaca, N.Y.: Cornell University Press, 1982), 287.
50. Jameson, *Archaeologies of the Future,* 175.
51. Ibid., 175–207.
52. Ibid., 211.
53. Jameson, *Postmodernism,* 39.
54. Ibid., 413.
55. Ibid., 31.
56. Jeffrey T. Nealon, *Post-Postmodernism: Or, the Cultural Logic of Just-in-Time Capitalism* (Stanford, Calif.: Stanford University Press, 2012), 7.
57. Wendy Hui Kyong Chun, *Programmed Visions: Software and Memory* (Cambridge, Mass.: MIT Press, 2011), 73.
58. Nealon, *Post-Postmodernism,* 31.
59. Franco "Bifo" Berardi, *The Soul at Work: From Alienation to Autonomy,* trans. Francesca Cadel and Giuseppina Mecchia (New York: Semiotex(e), 2009), 74.
60. FM-2030, *Are You a Transhuman? Monitoring and Stimulating Your Personal Rate of Growth in a Rapidly Changing World* (New York: Warner, 1989), n.p.
61. Ibid.
62. "About Us. Ray and Terry's Longevity Products," 2011, http://www.rayandterry.com/about-us/.
63. See, especially, his discussion of "a certain anarchism" (Hakim Bey's theory of the Temporary Autonomous Zone) in *Archaeologies of the Future* (Jameson, *Archaeologies of the Future,* 213). Despite offering a theory of a postmodern utopia that is, in essence, Bey's theory of the TAZ, Jameson in that section cannot acknowledge another revolutionary praxis as having a more valid approach than the orthodox Marxism to which he still retains strong political commitments. Thus, I think it is possible to see, in both the transhuman Jameson I am outlining and this anarchist Jameson in *Archaeologies of the Future,* the rich possibility for reading Jameson contra his own assertions about his work.

1. An Inner Transhumanism

1. Friedrich Nietzsche, *Thus Spoke Zarathustra,* trans. Adrian Del Caro (Cambridge: Cambridge University Press, 2006), 12.

2. Nick Bostrom, "A History of Transhumanist Thought," *Journal of Evolution & Technology* 14, no. 1 (2005), http://jetpress.org/.
3. Ibid.
4. Stefan Lorenz Sorgner, "Nietzsche, the Overhuman, and Transhumanism," *Journal of Evolution & Technology* 20, no. 1 (2009), http://jetpress.org/.
5. William Sims Bainbridge, "Burglarizing Nietzsche's Tomb," *Journal of Evolution & Technology* 21, no. 1 (2010), http://jetpress.org/.
6. Stefan Lorenz Sorgner, "Beyond Humanism: Reflections on Trans- and Posthumanism," *Journal of Evolution & Technology* 21, no. 2 (2010), http://jetpress.org/.
7. Max More, "The Overhuman in the Transhuman," *Journal of Evolution & Technology* 21, no. 1 (2010), http://jetpress.org/.
8. Sorgner, "Nietzsche, the Overhuman, and Transhumanism." n.p.
9. The other major modernist contributor to the origins of transhumanism is Henri Bergson with his concept of creative evolution (the idea that evolution *wants* something). I will be discussing Bergson through one of his most popular and successful transhuman appliers in chapter 3, Pierre Teilhard de Chardin; however, see Keith Ansell-Pearson, *Germinal Life: The Difference and Repetition of Deleuze* for an excellent account of Bergson and transhumanism. Ansell-Pearson's book purports to be about Deleuze and transhuman philosophy, but as the argument so closely hinges on Deleuze's readings of Bergson, in many ways the book can also be read as an account of Bergson and transhumanism.
10. Janet Lyon, *Manifestoes: Provocations of the Modern* (Ithaca, N.Y: Cornell University Press, 1999), 153.
11. Ibid., 150–51.
12. Ibid., 153–54.
13. Ibid., 154.
14. Ibid.,
15. Carolyn Burke, *Becoming Modern: The Life of Mina Loy* (Berkeley: University of California Press, 1997), viii.
16. Ibid., 78.
17. Michael Tanner, *Nietzsche: A Very Short Introduction* (Oxford: Oxford University Press, 2001), 30.
18. Ullrich Haase, *Starting with Nietzsche* (New York: Continuum, 2009), 153.
19. Ibid., 154.
20. I say "his" here because these types are always young men, which is, of course, the point of Loy's critique.
21. Lyon, *Manifestoes*, 154–155.
22. Ibid., 156–57.
23. Richard Maurice Bucke, *Cosmic Consciousness: A Study in the Evolution of the Human Mind.* (New York: E.P. Dutton, 1901), 1. Also available at http://www.sacred-texts.com/.
24. Ibid., 3.
25. Ibid., 3.

26. Ibid., 7.
27. Lyon, *Manifestoes,* 152.
28. Mina Loy, *The Lost Lunar Baedeker: Poems of Mina Loy,* ed. Roger L. Conover (New York: Farrar, Straus and Giroux, 1997), 151.
29. Filippo Tommaso Marinetti, *Marinetti: Selected Writings,* ed. R. W. Flint, trans. R. W. Flint and Arthur A. Coppotelli (New York: Farrar, Straus and Giroux, 1972), 41.
30. Ibid., 42.
31. Loy, *The Lost Lunar Baedeker,* 4.
32. Ibid., 6.
33. Stanislav Grof, *The Adventure of Self-Discovery: Dimensions of Consciousness and New Perspectives in Psychotherapy and Inner Exploration* (Buffalo: State University of New York Press, 1988), 30–34.
34. Loy, *The Lost Lunar Baedeker,* 6–7.
35. Ibid., 7.
36. Ibid.
37. Ibid.
38. R. Allen Shoaf, *Dante, Chaucer, and the Currency of the Word: Money, Images, and Reference in Late Medieval Poetry* (Norman, Okla.: Pilgrim Books, 1983), 21–25.
39. T. S. Eliot, *The Cocktail Party* (New York: Mariner Books, 1964), 147.
40. Alexander von Humboldt, *Cosmos: A Sketch of a Physical Description of the Universe,* trans. E. C. Otté (New York: Harper, 1860), 69.
41. Lyon, *Manifestoes,* 157.
42. Ibid., 157.
43. Ibid., 156.
44. Loy, *The Lost Lunar Baedeker,* 63.
45. Ibid., 65.
46. Nietzsche, *Thus Spoke Zarathustra,* 10.
47. Loy, *The Lost Lunar Baedeker,* 65.
48. Jeffey Meyers, *Katherine Mansfield: A Darker View* (New York: Cooper Square, 2002), 242.
49. P. D. Ouspensky, *A New Model of the Universe* (New York: Vintage, 1971) vii.
50. Ibid., 101.
51. Ibid.
52. Ibid.
53. Ibid., 102.
54. Ibid.
55. Ibid.
56. Ibid., 110.
57. Ibid., 103.
58. Ibid.
59. R. J. Hollingdale, *Nietzsche: The Man and His Philosophy* (Cambridge: Cambridge University Press, 2001), 72–73.
60. Ouspensky, *A New Model of the Universe,* 103.

61. Ibid., 110.
62. Ibid.
63. William A. Covino, "Magic and/as Rhetoric: Outlines of a History of Phantasy," *Journal of Advanced Composition* 12, no. 2 (October 1, 1992): 349.
64. Ibid.
65. Ibid.
66. Leigh Wilson, *Modernism and Magic: Experiments with Spiritualism, Theosophy, and the Occult* (Edinburgh: Edinburgh University Press, 2012), 15.
67. Ibid., 18.
68. Quoted. in ibid., 24.
69. Ibid.
70. Bergson's presidential address, "'Phantasms of the Living' and Psychical Research" has been published in the collection *Mind-Energy.* See Henri Bergson, *Mind-Energy* (New York: Palgrave, 2007).
71. Wilson, *Modernism and Magic,* 26.
72. George A. Kennedy, *Classical Rhetoric and Its Christian and Secular Tradition from Ancient to Modern Times,* 2nd ed. (Chapel Hill: University of North Carolina Press, 1999), 228.
73. Covino, "Magic and/as Rhetoric," 350.
74. Ibid., 352.
75. Ibid., 351–52.
76. "Current-traditional" is the term, often disparaging, for rhetoric textbooks used in the nineteenth and early twentieth centuries in American writing classrooms. These texts stressed final product over writing process and ignored the complicated and detailed process of composition outlined in Classical rhetoric.
77. Ibid., 351.
78. Ibid., 354.
79. Ouspensky, *A New Model of the Universe,* 101.
80. Ibid., 106–7.
81. Wilson, *Modernism and Magic,* 27.
82. Karl Popper, *Conjectures and Refutations: The Growth of Scientific Knowledge,* 2nd ed. (New York: Routledge, 2002), 48.
83. Kurzweil's date for this event has been around thirty years in the future since 1993 (which would have made the Singularity occur in 2023), though most of his recent predictions now situate the event in 2045. Ray Kurzweil, *The Singularity Is Near: When Humans Transcend Biology* (New York: Penguin, 2006), 135–36.
84. Ouspensky, *A New Model of the Universe,* 129.
85. Ibid., 118.
86. Ibid.
87. Ibid.
88. Ibid.
89. Ibid., 119.
90. Ibid., 110.

91. Ibid., 119.
92. Filippo Tommaso Marinetti, *Critical Writings: New Edition,* ed. Günter Berghaus, trans. Doug Thompson (New York: Farrar, Straus and Giroux, 2008), 86.
93. Bostrom, "A History of Transhumanist Thought," n.p.
94. Ibid.
95. Ibid.
96. Ouspensky, *A New Model of the Universe,* 108.
97. Richard Dawkins, *River out of Eden: A Darwinian View of Life* (New York: Basic, 1996), 121–22.
98. Ibid., 106.
99. Ibid., 121–22.
100. Elliott Sober, *Philosophy of Biology* (Boulder, Colo,: Westview, 2000), 91.
101. Charles Darwin, *The Descent of Man* (Amherst, N.Y.: Prometheus Books, 1997), 135.
102. Ibid., 137.
103. Ouspensky, *A New Model of the Universe,* 108.
104. Bernhard Rensch, *Homo Sapiens: From Man to Demigod,* trans. C. A. M. Sym (New York: Columbia University Press, 1972), 1.
105. Ibid., 3.
106. Ibid., 2.
107. Mina Loy, "History of Religion and Eros," in *Stories and Essays of Mina Loy,* ed. Sara Crangle (Champaign, Ill.: Dalkey Archive Press, 2011), 242.
108. Ouspensky, *A New Model of the Universe,* 111.
109. Loy, "History of Religion and Eros," 245.
110. Ibid.

2. Astounding Transhumanism!

1. Francis Fukuyama, "Transhumanism," *Foreign Policy,* September 1, 2004: 42.
2. Ibid.
3. Ibid.
4. Ibid.
5. Ibid.
6. Ibid.
7. James Hughes, "Introduction to the HETHR Papers," *Journal of Evolution & Technology* 18, no. 1 (2008), http://jetpress.org/.
8. Ibid.
9. Nick Bostrom, "A History of Transhumanist Thought," *Journal of Evolution & Technology* 14, no. 1 (2005), http://jetpress.org/.
10. Ibid.
11. Ibid.
12. Ibid.
13. Philip K. Dick, *The Shifting Realities of Philip K. Dick: Selected Literary and Philosophical Writings,* ed. Lawrence Sutin (New York: Vintage, 1996), 259.
14. This massive work represents Dick's attempt to work out an experience that oc-

curred in 1974. On February 3, 1974, Dick felt that he was contacted by a superhuman being, possibly an ancient artificial intelligence left by aliens in Earth orbit or merely God, and given a vision of the nature of the universe that conformed to classic insights from Gnosticism. He spent the final eight years of his life working through this experience. These explorations inspired his final novels: *Valis* (1978), *The Divine Invasion* (1980), *The Transmigration of Timothy Archer* (1982), and the posthumous *Radio Free Albemuth* (1985). They also resulted in the handwritten journal that is being transcribed by the Zebrapedia project at Penn State and selections of which have been published in Philip K. Dick, *The Exegesis of Philip K. Dick*, ed. Pamela Jackson and Jonathan Lethem (Boston: Houghton Mifflin Harcourt, 2011).

15. Dick, *The Shifting Realities of Philip K. Dick*, 259.
16. Ibid.
17. Ibid., x.
18. Ibid., 263.
19. Dick, *The Exegesis of Philip K. Dick*, xi.
20. Dick, *The Shifting Realities of Philip K. Dick*, 265.
21. Ibid., 266.
22. Ibid., 277.
23. Dick, *The Exegesis of Philip K. Dick*, 314.
24. Ibid., 97.
25. Ken MacLeod, Twitter Post, February 10, 2013, 4:24PM, http://twitter.com/amendlocke.
26. Dick, *The Exegesis of Philip K. Dick*, xvi.
27. H. Bruce Franklin, *Robert A. Heinlein: America as Science Fiction* (Oxford: Oxford University Press, 1980), 3.
28. Jack Parsons is himself a fascinating figure in both the early history of rocketry and the importation of British occultism into America. For more information, see George Pendle, *Strange Angel: The Otherworldly Life of Rocket Scientist John Whiteside Parsons* (New York: Mariner, 2006), and John Carter, *Sex and Rockets: The Occult World of Jack Parsons* (Port Townsend, Wash.: Feral House, 2005).
29. Anthony Boucher [H. H. Holmes], *Rocket to the Morgue* (New York: Duell, Sloan, Pearce, 1942), 192.
30. Thomas M. Disch, *The Dreams Our Stuff Is Made Of: How Science Fiction Conquered the World* (New York: Free Press, 1998), 117.
31. Ibid., 5.
32. Ibid., 6–7.
33. Ibid., 7.
34. William Gibson, *Burning Chrome* (New York: Ace, 1986), 27.
35. Andrew Ross, "Getting Out of the Gernsback Continuum," *Critical Inquiry* 17, no. 2 (January 1, 1991): 414.
36. Ibid.
37. Ibid., 430.
38. Ibid.

39. Ibid.
40. Ibid.
41. Dick, *The Shifting Realities of Philip K. Dick,* 264.
42. Bruce Sterling, "Preface," in *Burning Chrome,* by William Gibson (New York: Harper Voyager, 2003), xi.
43. Ross, "Getting Out of the Gernsback Continuum," 420.
44. It was published in hardback in 1946 by Arkham House, August Derleth and Donald Wandrei's press initially founded to keep in print the works of H. P. Lovecraft. Van Vogt also has the distinction of having the first SF hardback published by a major trade press when Ace published *The World of Null-A* in 1949.
45. A. E. van Vogt, *Reflections of A. E. van Vogt: The Autobiography of a Science Fiction Giant with a Complete Bibliography* (Lakemont, Ga.: Fictioneer, 1975), 83–84.
46. John W. Campbell, *Collected Editorials from Analog,* ed. Harry Harrison (New York: Doubleday, 1966), 217.
47. Ibid., 224.
48. Ibid., 223.
49. Ibid., 224.
50. Ibid.
51. For an account of how these associations play out later in Golden Age and New Wave science fiction, see N. Katherine Hayles, *How We Became Posthuman: Virtual Bodies in Cybernetics, Literature, and Informatics* (Chicago: University of Chicago Press, 1999).
52. Van Vogt, *Reflections,* 81.
53. Ibid., 92–93.
54. Ibid., 96.
55. Dick, *The Exegesis of Philip K. Dick,* 159–60.
56. Brian Attebery, "Super Men," *Science Fiction Studies* 25, no. 1 (March 1, 1998): 61–76.
57. Dick, *The Exegesis of Philip K. Dick,* 160.
58. A. E. van Vogt, *Slan* (New York: Orb, 1998), 166.
59. Ibid., 166, 185, 187.
60. Ibid., 186.
61. Disch, *The Dreams Our Stuff Is Made of,* 6.
62. Van Vogt, *Slan,* 191.
63. Attebery, "Super Men," 66.
64. Ibid., 64.
65. Van Vogt, *Slan,* 7.
66. I draw my inspiration for naming these communities as well as much of my thinking about them from Peter Lamborn Wilson's work on pirate utopias. See Peter Lamborn Wilson, *Pirate Utopias* (New York: Autonomedia, 2003).
67. The address is 25 Poplar St., Battle Creek, Michigan. The house still stands today.
68. Dal Coger, "The Legendary Slan Shack," *Mimosa,* no. 29 (December 2002), http://jophan.org/.
69. "Slan Shack," in *Fancyclopedia 3,* 2013, http://fancyclopedia.org. The names of

some of these dwellings, as documented in 1959's *Fancyclopedia 2,* are Tendril Tower in Los Angeles, The Ivory Birdbath in Cambridge, Massachusetts, Granny's House in Washington, D.C., as well as The Futurian House, The Futurian Fortress, The Futurian Embassy, Ivory Tower, Prime Base, Raven's Roost, Station X, and The Hatch—all in New York City.

70. Coger, "The Legendary Slan Shack," n.p.
71. "Numerical Fandoms," in *Fancyclopedia 3,* 1959, http://fancyclopedia.org/. Although, as documented in *Fancyclopedia 2,* the numbering would continue, with the 1940s closing in the middle of "Fifth Fandom."
72. Harry Warner Jr., "Al Ashley," *All Our Yesterdays,* no. 21 (2000), http://efanzines.com.
73. Ibid. n.p.
74. Curt Phillips, "One Life, Furnished in Early Fandom," *Mimosa,* no. 30 (August 2003), http://jophan.org n.p.
75. Ibid.
76. Coger, "The Legendary Slan Shack," n.p.
77. Warner Jr., "Al Ashley," n.p.
78. Al Ashley and Abby Lu Ashley, "Slan Center," *En Garde* 6 (June 1943). Coslet Collection. Special Collections, University of Maryland, Baltimore County: 2.
79. Ibid., 4.
80. Ibid., 3–4.
81. Ibid., 2.
82. Ibid., 1.
83. Ibid., 4.
84. Claude Degler, "Cosmic Circle Commentator," *Cosmic Circle Commentator* 1 (September 1943). Coslet Collection. Special Collections, University of Maryland, Baltimore County: 1.
85. Ibid., 3. All quotations from Claude Degler's mimeographed fanzines have emphasis added in the original. As noted in his entry in *Fancylcopedia 2,* "the most noticeable characteristic" of Degler's "publications was that they were the worst-looking legible fanzines ever published; abounding strikeovers, paragraphs non-existent, stencils crowded to the edges, no spacing after periods, misspelling, overuse of capitals, quotemarks and underlines, wandering unplanned sentences, grammatical errors like 'can and has went,' malapropisms like calling Widner a stolid and far-seeing fan, ad nauseam." "Claude Degler," in *Fancyclopedia 3,* 1959, http://fancyclopedia.org/ n.p.
86. Degler, "Cosmic Circle Commentator," 2.
87. Claude Degler, "Cosmic Circle Monthly," *Cosmic Circle Commentator* 1 (July 1944). Coslet Collection. Special Collections, University of Maryland, Baltimore County: 4.
88. Degler, "Cosmic Circle Commentator," 2.
89. Warner Jr., "Al Ashley," n.p. This settlement, in a move that is probably not coincidental, furnishes almost exactly the plot of John Varley's 1978 story "The Persistence of Vision," one of SF's most famous stories.

90. There do not appear to be any surviving copies of this journal. Given Degler's anticipatory style, in which, for instance, he mentions "two books about *fandom* and the *Cosmic Circle* in the world of the future are now in preparation" that are not known to have ever been published or even written, as well as the constantly changing titles of his mimeographed fanzines, there may not have ever been a *Cosmic Circle Digest* (Degler, "Cosmic Circle Commentator," 2).
91. Ibid., 4.
92. Ibid., 5.
93. "Claude Degler," n.p.
94. "Numerical Fandoms" n.p.
95. Ashley and Ashley, "Slan Center," 4.
96. Richard A. Lanham, *Motives of Eloquence: Literary Rhetoric in the Renaissance* (New Haven, Conn.: Yale University Press, 1976), 1.
97. Ibid., 6.
98. Ibid., 4.
99. Ibid., 7. Lanham uses "centermentalism" to designate the manner in which philosophy focuses on the *pathos* of human existence (our inevitable death in a seemingly uncaring universe). Figuring existence as pathetic has the effect of situating our struggle to find our way in the world as *the* central struggle and us as the center of our universe. In other words, rhetoric's move beyond meaning is crucially a move beyond solipsism.
100. Ibid., 4.
101. Ibid.
102. Lanham uses "discovering" in rather strongly implied quotations, as he uses "discovering" in the same way as Nietzsche's famous definition of truth: "If someone hides something behind a bush, looks for it in the same place and then finds it there, his seeking and finding is nothing much to boast about." Friedrich Nietzsche, "On Truth and Lying in a Non-Moral Sense," in *The Birth of Tragedy and Other Writings*, ed. Raymond Geuss and Roland Speirs, trans. Roland Speirs (Cambridge: Cambridge University Press, 1999), 147. Also, Stanley Fish in his commentary on Lanham's distinction in "Rhetoric" connects Lanham's concept of rhetorical man, through Paul De Man, to the Nietzsche of "On Truth and Lying in a Non-Moral Sense." See Stanley Fish, "Rhetoric," in *Doing What Comes Naturally: Change, Rhetoric, and the Practice of Theory in Literary and Legal Studies* (Durham, N.C.: Duke University Press, 1989), 471–502.
103. Lee Edelman, *No Future: Queer Theory and the Death Drive* (Durham, N.C.: Duke University Press, 2004), 2.
104. Ibid., 2–3.
105. Ibid., 151.
106. Fredric Jameson, *Archaeologies of the Future: The Desire Called Science Fiction and Other Science Fictions* (London: Verso, 2005), 211.
107. I don't mean to imply that queer futurity is not serious, but, following Edelman's logic through, the kinds of bioconservative arguments the transhumanists grapple with are of an order similar to those arguments that privilege children above all

else in Western culture and to those that attempt to exclude figures such as Claude Degler.

108. Claudia Crawford, "'The Dionysian Worldview': Nietzsche's Symbolic Languages and Music," *Journal of Nietzsche Studies,* no. 13 (April 1, 1997): 72.
109. Friedrich Nietzsche, "The Dionysiac Worldview," in *The Birth of Tragedy and Other Writings,* ed. Raymond Geuss and Roland Speirs, trans. Roland Speirs (Cambridge: Cambridge University Press, 1999), 120.
110. Ibid.
111. Ibid.
112. Ibid.
113. Ibid.
114. Ibid., 138.
115. Ibid., 121.

3. Toward Omega

1. For the fuller version of this project, see Isabelle Stengers, *Cosmopolitics I,* trans. Robert Bononno (Minneapolis: University of Minnesota Press, 2010) and Isabelle Stengers, *Cosmopolitics II,* trans. Robert Bononno (Minneapolis: University of Minnesota Press, 2011).
2. Isabelle Stengers, "The Cosmopolitical Proposal," in *Making Things Public: Atmospheres of Democracy,* ed. Bruno Latour and Peter Weibel (Cambridge, Mass.: MIT Press, 2005), 994.
3. Ibid., 995.
4. Ibid.
5. Ibid., 999.
6. Ibid.
7. Paul Virilio, *Strategy of Deception* (London: Verso, 2007), 8.
8. Paul Virilio and Sylvère Lotringer, *Pure War,* trans. Mark Polizzoitti and Brian O'Keeffe (Los Angeles: Semiotext(e), 2008), 73.
9. Charles Stross, *Singularity Sky* (New York: Ace, 2004), 143.
10. Ibid., 252–53.
11. Although I will not be addressing the obvious critique that might result from considering this movement with the prior, same-named movement focused on freeing slaves and ending slavery throughout the world in the nineteenth century, there seems to be a rather disturbing intertextual reading of the two movements: one focused on ending one of history's greatest evils and the other on the end of pain. Generally, Pearce's choice of name probably is intended to signal another kind of emancipation, from pain instead of bondage, but the naming equally highlights the generally tone-deaf nature of contemporary transhumanism on issues of race and gender.
12. Robert Ettinger, *Man into Superman: The Startling Potential for Human Evolution—and How to Be a Part of It* (Clinton Township, Mich.: Cryogenics Institute, 1989), 9, http://www.cryonics.org/images/uploads/misc/ManIntoSuperman.pdf.

13. David Pearce, "0.1," *The Hedonistic Imperative,* 1997, http://www.hedweb.com/. *The Hedonistic Imperative* is available as a web text. As such, citations are to the chapter and section number, as presented on Pearce's web version of the work, http://hedweb.org
14. Ibid., 0.1.
15. Ibid., 0.3.
16. Ibid., 0.1.
17. Ibid. abstract.
18. Ibid., 0.3.
19. Ibid.
20. Ibid., 0.4.
21. Ibid., 2.0.
22. David Pearce, "Top Five Reasons Transhumanism Can Eliminate Suffering," *H+* (2012), http://hplusmagazine.com/.
23. David Pearce, "The Abolitionist Project," 2007, http://www.abolitionist.com/.
24. Stengers, "The Cosmopolitical Proposal," 999. Juxtaposing Stengers and Pearce is not meant to suggest that Pearce does not care about these kinds of questions. Much of his work is taken up with thinking through the implications of a postsuffering heaven-on-Earth; however, as with the discussion of the wired-up rat and the subtler version we must seek, the idea that we might get paradise-engineering wrong or that pain may actually be a more helpful motive than pleasure are not considered in his work. This unwillingness is, in my reading of Stengers, the very definition of the blindness of the entrepreneur.
25. Jean-Francois Lyotard, *The Inhuman: Reflections on Time,* trans. Geoffrey Bennington and Rachel Bowlby (Palo Alto, Calif.: Stanford University Press, 1992), 9.
26. Ibid., 11. This argument parallels Quentin Meillassoux's usage of what he calls "ancestrality" to reject the Kantian philosophical correlate in which it is impossible to think either the subject or the object outside of their relationship. He writes of ancestrality—the fossil record that proves a world existed prior to thought—"that the singular birth and death of consciousnesses implies a time which cannot itself be of the order of consciousness." Quentin Meillassoux, *After Finitude: An Essay on the Necessity of Contingency* (New York: Bloomsbury, 2009), 130. Ray Brassier picks up on this element of Lyotard in "Solar Catastrophe: Lyotard, Freud, and the Death-Drive," *Philosophy Today* (January 1, 2003), and intensifies it in a revision of that article in Ray Brassier, *Nihil Unbound: Enlightenment and Extinction* (New York: Palgrave Macmillan, 2010). In the 2003 article, Brassier focuses more on the Freudian elements of Lyotard's discussion, while the 2010 revision connects more directly to Meillassoux's attack on post-Kantian correlationism. Both, however, ignore Lyotard's extensive engagement with extropian transhumanism.
27. Lyotard, *The Inhuman,* 13.
28. Ibid., 13.
29. Ibid., 14.
30. Ibid., 17.
31. Ibid.

32. For perhaps the best treatment of the issue of whether an uploaded consciousness is fully human or *merely* a simulation, see William Gibson, "Winter Market," in *Burning Chrome* (New York: Ace, 1986), 124–51. In that story, a VR producer and his artist friend stage a series of debates about whether a recently uploaded disabled colleague will be fully human inside of the computer or will only merely be a program that pretends to be their now-dead friend.
33. Lyotard, *The Inhuman*, 17.
34. Ibid.
35. Ibid.
36. Ibid., 18.
37. Ibid., 19.
38. Ibid.
39. Ibid., 20.
40. Friedrich Nietzsche, "On Truth and Lying in a Non-moral Sense," in *The Birth of Tragedy and Other Writings*, ed. Raymond Geuss and Roland Speirs, trans. Roland Speirs (Cambridge: Cambridge University Press, 1999), 153.
41. Ibid., 153.
42. Frank J. Tipler, *The Physics of Immortality: Modern Cosmology, God, and the Resurrection of the Dead* (New York: Anchor, 1997), 56–57.
43. Ibid., 57.
44. This view of human intelligence (and all life, for that matter) is borne out in the history of information theory. For instance, Norbert Wiener, while introducing his groundbreaking work on cybernetics to the general public in *The Human Uses of Human Beings*, refers to humans as "pockets of decreasing entropy in a framework in which the large entropy tends to increase." Norbert Wiener, *The Human Use of Human Beings: Cybernetics and Society* (New York: Da Capo, 1988), 32.
45. Tipler, *The Physics of Immortality*, 64.
46. See Henri Bergson, *Creative Evolution*, ed. K. Ansell-Pearson, M. Kolkman, and M. Vaughan (New York: Palgrave Macmillan, 2007).
47. Marco Bischof, "Vernadsky's Noosphere and Slavophile Sobornost'," in *Biophotonics and Coherent Systems in Biology*, ed. L. V. Beloussov, V. L. Voeikov, and V. S. Martynyuk (Boston, Mass.: Springer, 2007), 280.
48. Ibid., 280.
49. Vaclav Smil, *The Earth's Biosphere: Evolution, Dynamics, and Change* (Cambridge, Mass.: MIT Press, 2003), 13.
50. Vladimir I. Vernadsky, "The Biosphere and the Noösphere," *Executive Intelligence Review* (February 18, 2005).
51. Bischof, "Vernadsky's Noösphere and Slavophile Sobornost'," 281–82.
52. Pierre Teilhard de Chardin, *Human Energy* (New York: Harper, 1969), 145.
53. This hybridity is especially evident in *The Phenomenon of Man*, Teilhard's most widely read work. However, many who reject Teilhard for his mysticism and his lack of precision should refer to the more recognizably philosophical or rational tone of essay collections such as *The Future of Man* or *Activation of Energy*.

54. Nick Bostrom, "A History of Transhumanist Thought," *Journal of Evolution & Technology* 14, no. 1 (2005), http://jetpress.org/volume14/bostrom.html.
55. In a 1965 paper, Gordon E. Moore, a founder of Intel, claimed that the number of transistors that could fit onto a fixed area of silicon would double every year. This trend held until starting to slow in 2016 and has resulted in a continued, upward rise in available computing power.
56. Vernor Vinge, "The Coming Technological Singularity: How to Survive in the Post-Human Era," in *VISION-21 Symposium Sponsored by NASA Lewis Research Center and the Ohio Aerospace Institute,* 1993, http://www.rohan.sdsu.edu/faculty/vinge/.
57. Ibid.
58. Ibid.
59. Bostrom, "A History of Transhumanist Thought."
60. Bertrand Russell, *History of Western Philosophy* (London: Allen And Unwin, 1946), 827.
61. For an overview of the Einstein–Bergson debate, see Jimena Canales, "Einstein, Bergson, and the Experiment That Failed: Intellectual Cooperation at the League of Nations," *MLN* 120, no. 5 (2005): 1168–91.
62. Pierre Teilhard de Chardin, *The Phenomenon of Man* (New York: Harper, 1965), 29.
63. Ibid.
64. Ibid., 26.
65. K. Eric Drexler, *Radical Abundance: How a Revolution in Nanotechnology Will Change Civilization* (New York: PublicAffairs, 2013), 113.
66. Ibid., 117–18.
67. Bostrom, "A History of Transhumanist Thought."
68. Teilhard de Chardin, *The Phenomenon of Man,* 31.
69. Richard Doyle, *Darwin's Pharmacy: Sex, Plants, and the Evolution of the Noösphere* (Seattle: University of Washington Press, 2011).
70. Teilhard de Chardin, *The Phenomenon of Man,* 180–81.
71. *Le Phénomene Humain,* the work's original French title, is more accurately translated as "the human phenomenon," a closer approximation of Teilhard's meaning.
72. Pierre Teilhard de Chardin, *Writings in a Time of War* (New York: Harper, 1967), 18.
73. Teilhard de Chardin, *The Phenomenon of Man,* 32.
74. Ibid., 32.
75. Ibid., 34.
76. Ibid., 240.
77. Teilhard de Chardin, *Human Energy,* 145.
78. Teilhard de Chardin, *The Phenomenon of Man,* 19.
79. Ibid.
80. Ibid.
81. Ibid.
82. Ibid., 258.
83. Ibid., 259.

84. Julian Huxley, "Transhumanism," in *New Bottles for New Wine* (New York: Harper, 1957), 14.
85. Ibid., 14–15.
86. Ibid., 15.
87. Ibid., 13–14.
88. Timothy Leary, "The Cyber-Punk: The Individual as Reality Pilot," *Mississippi Review* 16, no. 2 (January 1, 1988): 252.
89. Ibid., 253.
90. Ibid., 254.
91. Ibid.
92. Ibid., 256.
93. Huxley, "Transhumanism," 13.
94. Teilhard de Chardin, *The Phenomenon of Man*, 244–45.
95. Ibid., 245.
96. Ibid., 313.
97. Teilhard de Chardin, *Writings in a Time of War*, 13.
98. Ibid., 67.
99. Ibid., 66.
100. Ibid.
101. Pierre Teilhard de Chardin, *The Future of Man* (New York: Harper, 1964), 75.
102. Teilhard de Chardin, *Writings in a Time of War*, 57–58.
103. Ibid., 59.
104. Stengers, "The Cosmopolitical Proposal," 995.
105. Teilhard de Chardin, *Writings in a Time of War*, 66.

4. Transhuman Aesthetics

1. Nick Bostrom, "A History of Transhumanist Thought," *Journal of Evolution & Technology* 14, no. 1 (2005), http://jetpress.org/volume14/.
2. Natasha Vita-More, "Transhuman Manifesto," 1983, http://transhumanist.biz/. Both versions of the manifesto have disappeared from the Internet but are accessible through Archive.org's Wayback Machine.
3. Natasha Vita-More, "Transhumanism Art Manifesto," 2003, http://www.transhumanist.biz/ n.p.
4. Shusaku Arakawa and Madeline Gins, *Architectural Body* (Tuscaloosa: Alabama UP, 2002) xii.
5. Shusaku Arakawa and Madeline Gins, *Reversible Destiny* (New York: Abrams, 1997), 313.
6. Ibid., 313.
7. Arakawa and Gins, *Architectural Body*, xvi.
8. Ibid., xv–xvi.
9. Ibid., xx.
10. Ibid., xviii.
11. Ibid., xx.

12. Ibid., xxi.
13. Ibid., 1.
14. Arakawa and Gins, *Reversible Destiny,* 151.
15. Ibid., 12.
16. Ibid., 154.
17. Arakawa and Gins, *Architectural Body,* xx.
18. Ibid., 23, 25.
19. Ibid., 29.
20. Ibid., 1.
21. Ibid., 97.
22. http://www.huffingtonpost.com/zoltan-istvan/transhumanist-art-is-brin_b_5447758.html.
23. Zoltan Istvan, "Transhumanist Art Will Help Guide People to Becoming Masterpieces," *Huffington Post,* June 5, 2014, http://www.huffingtonpost.com/zoltan-istvan/.
24. Ettinger, *Man into Superman,* 9.
25. Ibid.
26. Istvan, "Transhumanist Art Will Help Guide People to Becoming Masterpieces," n.p.
27. Tommaso Toffoli and Norman Margolus, "Programmable Matter: Concepts and Realization," *Physica D: Nonlinear Phenomena* 47, no. 1 (January 1, 1991): 263–72.
28. Ibid., 263.
29. Kurzweil, *The Singularity Is Near,* 131.
30. Ibid.
31. Ibid., 137.
32. Martin Heidegger, *Basic Writings: From 'Being and Time" (1927) to "The Task of Thinking" (1964)* (New York: HarperCollins, 1993), 322.
33. Kurzweil, *The Singularity Is Near,* 137.
34. Edmund Burke, *A Philosophical Enquiry into the Origins of the Sublime and Beautiful: And Other Pre-Revolutionary Writings,* ed. David Womersley (New York: Penguin, 1999), 80.
35. Ibid., 108.
36. Ibid., 106.
37. Ibid., 112.
38. Gilles Deleuze and Félix Guattari, *What Is Philosophy?* (New York: Columbia University Press, 1996), 212.
39. Ibid., 212–13.
40. Joel McKim and Brian Massumi, "Of Microperception and Micropolitics: An Interview with Brian Massumi," *INFLeXions* 3 (2008), http://www.senselab.ca/inflexions/volume_3/node_i3/massumi_en_inflexions_vol03.html.
41. Deleuze and Guattari, *What Is Philosophy?* 210.
42. Ibid., 210.
43. Ibid., 213.
44. Ibid., 203.

45. James Bridle, "The New Aesthetic," *Really Interesting Group* (May 6, 2011), http://www.riglondon.com/blog/.
46. http://www.comedycentral.com/.
47. Ibid.
48. Kyle McDonald, "Personifying Machines, Machining Persons," in *In Response to Bruce Sterling's "Essay on the New Aesthetic,"* ed. Julia Kaganskiy, 2012, http://www.thecreatorsproject.com/blog/.
49. Bruce Sterling, "An Essay on the New Aesthetic," *Beyond the Beyond* (April 2, 2012), http://www.wired.com/beyond_the_beyond/.
50. Ibid.
51. Ibid.
52. Marius Watz, "The Problem with Perpetual Newness," in *In Response to Bruce Sterling's "Essay on the New Aesthetic,"* ed. Julia Kaganskiy, 2012, http://www.thecreatorsproject.com/blog/.
53. Ibid. The mention of the sign is in reference to the number of signs, primarily in countries such as China that use a non-Latin character set, that people translate using a service like Google Translator and do not realize that their English "translation" is actually a report of an error. There are a surprisingly large number of these occurrences and numerous blogs dedicated to documenting them.
54. Kyle Chayka, "The New Aesthetic: Going Native," in *In Response to Bruce Sterling's "Essay on the New Aesthetic,"* ed. Julia Kaganskiy, 2012, http://www.thecreatorsproject.com/blog/.
55. Hal. Foster, "Prosthetic Gods," *Modernism/Modernity* 4, no. 2 (1997): 5.
56. Ibid., 7; emphasis in original.
57. Ibid., 9.
58. McDonald, "Personifying Machines, Machining Persons."
59. James Bridle, "Waving at the Machines," in *Web Directions South* (presented at the Web Directions South Conference, Sydney, Australia, 2011), http://www.webdirections.org/.
60. Terry Bisson, "The Singularity," *Postscripts* no. 8 (October 2006), http://www.terrybisson.com/.
61. Ibid.
62. Ibid.
63. Vaclav Smil, *Transforming the Twentieth Century: Technical Innovations and Their Consequences* (Oxford: Oxford University Press, 2006), 131.
64. Bisson, "The Singularity."
65. Bridle, "Waving at the Machines."
66. Sterling, "An Essay on the New Aesthetic."
67. Second only to Comic Sans MS as the most widely reviled typeface.
68. "Image Macro," *Wikipedia*, 2010, http://en.wikipedia.org/ n.p.
69. Ibid. n.p.
70. Lev Grossman, "Lolcats Addendum: Where I Got the Story Wrong," *TimeTechland* (July 16, 2007), http://techland.com/.

71. Lev Grossman, "Creating a Cute Cat Frenzy," *Time,* July 12, 2007, n.p.
72. See http://knowyourmeme.com/ for more information on this phenomenon.
73. Roland Barthes, *Mythologies,* trans. Annette Lavers (New York: Farrar, Straus & Giroux, 1972), 111.
74. Peter Russell, *The Global Brain: Speculations on the Evolutionary Leap to Planetary Consciousness* (New York: Tarcher, 1983), 111.
75. Ibid.
76. H. G. Wells, *World Brain: The Idea of a Permanent World Encyclopaedia,* 1937, https://sherlock.ischool.berkeley.edu/wells/.
77. Arthur C. Clarke, *The Collected Stories of Arthur C. Clarke* (New York: Macmillan, 2002), 822.
78. Ibid., 826.
79. Visit http://knowyourmeme.com/memes/tenso for more information on this meme.
80. "Teh" attains a particularly important meaning in meme culture because it is a very frequent typo in English that is tied to handedness.
81. Tim O'Reilly's famous overview ("Web 2.0 Compact Definition: Trying Again") argues for this understanding. In his account, Web 2.0 emerges as companies figured out how to build businesses with "the internet as platform" that work with "network effects" to imagine new forms of commerce. Tim O'Reilly, "Web 2.0 Compact Definition: Trying Again," *O'Reilly Radar* (December 10, 2006), http://radar.oreilly.com/ n.p.
82. Ibid. n.p.
83. Fredric Jameson, *The Hegel Variations: On the Phenomenology of the Spirit* (London: Verso, 2010), 4.

Conclusion

1. Benjamin Noys, *The Persistence of the Negative: A Critique of Contemporary Continental Theory* (Edinburgh: Edinburgh University Press, 2012), 4.
2. Ibid., 5.
3. Ibid., 8.
4. Ibid.
5. Jean-François Lyotard, *Peregrinations: Law, Form, Event,* The Wellek Library Lectures at the University of California, Irvine (New York: Columbia University Press, 1988), 13.
6. Benjamin Noys, *Malign Velocities: Accelerationism and Capitalism* (Winchester, U.K.: Zero Books, 2014) x.
7. McKenzie Wark, *Molecular Red: Theory for the Anthropocene* (London: Verso, 2015), 180.
8. F. M. Esfandiary, *Optimism One* (New York: Popular Library, 1978), 13.
9. Ibid., 14.
10. Ibid., 41.
11. F. M. Esfandiary, *Up-Wingers* (New York: John Day, 1973), ix–x.
12. Ibid., 84.

13. Ibid., 85.
14. Esfandiary, *Optimism One,* 45.
15. Ibid., 13.
16. Ibid., 41.
17. Esfandiary, *Up-Wingers,* 28.
18. Nick Land, *Fanged Noumena: Collected Writings 1987–2007,* ed. Ray Brassier and Robin Mackay (New York: Sequence, 2011), 35.
19. Ibid., 32.
20. Nick Land, "Critique of Transcendental Miserablism," in *Fanged Noumena: Collected Writings 1987–2007,* ed. Ray Brassier and Robin Mackay (New York: Sequence, 2011), 626.
21. Esfandiary, *Up-Wingers,* 85.
22. William Gibson, *Neuromancer* (New York: Ace, 1984), 1.
23. Esfandiary, *Up-Wingers,* 11.
24. Esfandiary, *Optimism One,* 22.
25. Ibid., 13.
26. Donna J. Haraway, "Cyborgs to Companion Species: Reconfiguring Kinship in Technoscience," in *The Haraway Reader* (New York: Routledge, 2003), 302–3.
27. Donna J. Haraway, *When Species Meet* (Minneapolis: University of Minnesota Press, 2007), 135.
28. Ibid., 135–36.
29. Ibid.
30. Esfandiary, *Optimism One,* 41.
31. Wark, *Molecular Red,* 120.
32. Ibid.
33. Ibid.
34. For examples of this crisis in political theory, see Franco "Bifo" Berardi, *After the Future,* ed. Gary Genosko and Nicholas Thoburn (Oakland, Calif.: AK Press, 2011). For an example of this crisis in science fiction, see Neal Stephenson, "Innovation Starvation," Wired October 27, 2011, http://www.wired.com, and the stories published in response in Ed Finn and Kathryn Cramer, *Hieroglyph: Stories and Visions for a Better Future* (New York: William Morrow, 2015).
35. Alex Williams and Nick Srnicek, "#Accelerate: Manifesto for an Accelerationist Politics," in *#Accelerate: The Accelerationist Reader,* ed. Robin Mackay and Armen Avanessian (Falmouth, U.K.: Urbanomic, 2014), 349.
36. Ibid. 351.
37. Ibid.
38. Ibid., 352.
39. Ibid.
40. Ibid.
41. Ibid., 353.
42. Ibid., 354.
43. Ibid.
44. Ibid., 355.

45. Ibid., 354.
46. Ibid., 355.
47. Ibid.
48. Ibid.
49. Ibid., 361.
50. Ibid., 362.
51. Ibid.
52. Emma E. Wilson, "Cyborg Anamnesis: #Accelerate's Feminist Prototypes," *PLATFORM: Journal of Media & Communication* 6, no. 2 (July 2015): 33–45.
53. "Xenofeminist Manifesto" exists as a hypertext hosted at http://laboriacuboniks.net and as a combined essay on a variety of other sites. However, as the manifesto has not yet been published in English in a print edition, I will be citing it using the hexadecimal section addresses that number the original text. Parenthetical citations in this section refer to Laboria Cuboniks, "Xenofeminism: A Politics for Alienation," 2015, http://laboriacuboniks.net.
54. Peter L. Berger and Thomas Luckmann, *The Social Construction of Reality: A Treatise in the Sociology of Knowledge* (New York: Anchor, 1967), 89.
55. Ibid.
56. Ibid.
57. Paul B. Preciado, *Testo Junkie: Sex, Drugs, and Biopolitics in the Pharmacopornographic Era* (New York: The Feminist Press, 2013).
58. For an excellent summary of the dialogue as a failed attempt by a master philosopher to engage talented rhetoricians in dialogue, while lacking the ability to persuade them to participate, see James L. Kastely, "In Defense of Plato's *Gorgias*," *PMLA* 106, no. 1 (1991): 96–109.
59. Plato, *Gorgias*, trans. Donald J. Zeyl (Indianapolis: Hackett, 1987), 512d–e.
60. These impressions of truth are *eidos* in Greek. For the Socrates depicted in Plato's dialogues, they make up the foundation of a true philosophy of the just and the good. This is in contrast to mere opinion (*doxa* in Greek), which is the subject of the flattery of the soul performed by orators like Callicles.
61. Plato, *Gorgias*, 501a–b.
62. This is Zeyl's translation of the word; the Greek here is often translated as "cookery," but as Zeyl clarifies, cooking is not Socrates's meaning here, as cooking had medicinal value for the Greeks. Pastry baking is more akin to the kind of decadent, empty calories Socrates is worried has fed the souls of Athens's citizens for too long.
63. Ibid., 518e.
64. Williams and Srnicek, "#Accelerate," 360–61.
65. For an account of Project Cybersyn, see Eden Medina, *Cybernetic Revolutionaries: Technology and Politics in Allende's Chile* (Cambridge, Mass.: MIT Press, 2014).
66. Cuboniks, "Xenofeminism," 0x0D.
67. Delphi Carstens, "Hyperstition. Xenopraxis," 2013, http://xenopraxis.net/.
68. Ibid. n.p.

69. Plato, *Gorgias,* 527d–e.
70. Isaiah Berlin, *Liberty: Incorporating Four Essays on Liberty* (Oxford: Oxford University Press, 2002), 301.
71. Ibid., 299.
72. Ibid., 301.
73. Plato, *Gorgias,* 491d.

INDEX

ANDREW PILSCH is assistant professor of English at Texas A&M University.

CPSIA information can be obtained
at www.ICGtesting.com
Printed in the USA
LVOW04*0542210817
545418LV00005BA/10/P